Palgrave Studies in Education and the Environment

Series Editors
Alan Reid
Faculty of Education
Monash University
Victoria, Victoria
Australia

Marcia McKenzie
College of Education
University of Saskatchewan
Saskatoon, Saskatchewan
Canada

This series focuses on new developments in the study of education and environment. Promoting theoretically-rich works, contributions include empirical and conceptual studies that advance critical analysis in environmental education and related fields. Concerned with the underlying assumptions and limitations of current educational theories in conceptualizing environmental and sustainability education, the series highlights works of theoretical depth and sophistication, accessibility and applicability, and with critical orientations to matters of public concern. It engages interdisciplinary and diverse perspectives as these relate to domains of policy, practice, and research. Studies in the series may span a range of scales from the more micro level of empirical thick description to macro conceptual analyses, highlighting current and upcoming turns in theoretical thought. Tapping into a growing body of theoretical scholarship in this domain, the series provides a venue for examining and expanding theorizations and approaches to the interdisciplinary intersections of environment and education. Its timeliness is clear as education becomes a key mode of response to environmental and sustainability issues internationally. The series will offer fresh perspectives on a range of topics such as:

- curricular responses to contemporary accounts of human-environment relations (e.g., the Anthropocene, nature-culture, animal studies, transdisciplinary studies)
- the power and limits of new materialist perspectives for philosophies of education
- denial and other responses to climate change in education practice and theory
- place-based and land-based orientations to education and scholarship
- postcolonial and intersectional critiques of environmental education and its research
- policy research, horizons, and contexts in environmental and sustainability education

More information about this series at
http://www.springer.com/series/15084

Bob Jickling • Stephen Sterling
Editors

Post-Sustainability and Environmental Education

Remaking Education for the Future

Foreword by David W. Orr

Editors
Bob Jickling
Lakehead Univeristy
Thunder Bay, Ontario, Canada

Stephen Sterling
Centre for Sustainable Futures (CSF)
University of Plymouth
Plymouth, UK

Palgrave Studies in Education and the Environment
ISBN 978-3-319-51321-8   ISBN 978-3-319-51322-5 (eBook)
DOI 10.1007/978-3-319-51322-5

Library of Congress Control Number: 2017932108

Cover illustration: Pattern adapted from an Indian cotton print produced in the 19th century

Printed on acid-free paper

This Palgrave Macmillan imprint is published by Springer Nature
The registered company is Springer International Publishing AG
The registered company address is: Gewerbestrasse 11, 6330 Cham, Switzerland

SERIES INTRODUCTION

Our primary goal for the *Palgrave Studies in Education and the Environment* series is to showcase new developments and advances in the scholarship of education and environment.

A key dimension of this aim is to promote theoretically rich work through contributions that include empirical and conceptual studies progressing critical analysis and practice in environmental education and related fields. In other words, as with our publishers, we expect the series to realize two outcomes: (i) advance the theoretical depth and sophistication of scholarship on education and environment, and (ii) offer critical orientations to such matters of public concern.

Why have we developed such expectations for this series?

First, there is our experience and sense of the strengths and weaknesses of existing scholarship in this area, echoed in the comments of our colleagues, mentors, and students. These raise a critical question for us: whether some of the underlying orientations of current and prevailing ways of conceptualizing and enacting environmental and sustainability education are *fundamentally limited* and *need shaking up*. This impetus pertains to both outcomes identified above, in that there is scope for broader and deeper theoretical engagement, as well as further consideration of the real implications of the scholarship for education and the environment. The series thus aims to highlight and support critical and theoretical scholarship that matters for how we live and educate in the world.

Second, to address such concerns, the series should enable readers to engage interdisciplinary and diverse perspectives on education and environment, particularly as these relate to domains of policy, practice, and research. Thus, we expect studies in this series to span a range of traditions, scales, and approaches, from the micro level of empirical thick description to the meta level of conceptual analysis and synthesis. Critically engaging with contemporary topics and issues demands high quality contributions that also both tap into a growing body of theoretical scholarship relevant to education and environment, and innovate in this space.

The series thus provides established and new scholars with both a venue and an avenue for examining the interdisciplinary intersections of environment and education, and challenging the theorizations and enactment of environment and sustainability-related education through critical, creative, and compelling scholarship.

We hope you enjoy engaging with the study that follows, and find it a fitting contribution to the series.

Alan Reid and Marcia McKenzie
Series Editors, *Palgrave Studies in Education and the Environment*

FOREWORD

The realities are these. Those of us who teach environment-related subjects, in various departments, are mostly employed in large organizations that are not wholly supportive or understanding of what we do or why we do it. For the most part we are tolerated—not always and everywhere, mind you, but mostly. We exist as outliers—a curricular out-shed behind the big house where the really important stuff happens. The reasons are many but I think they all reflect the failure of systems thinking throughout institutions of higher education. There can be no serious discussion about any environmental topic without understanding the larger system in which it is a part. In short, systems thinking is the study of what's hitched to what over what periods of time. The word "system," I think, is the most radical word in any language because it implies implicatedness—ecological, moral, political, economic—between what otherwise appear to be unrelated phenomena. Furthermore, since we cannot know the full extent of what causes what in complex, interactive, nonlinear systems especially over long periods of time, systems thinking begins in deep humility, not as a pose or gesture, but as an honest acknowledgement of our inescapable ignorance. But the humility required to acknowledge interrelatedness and its consequences is not well regarded in rigidly structured institutions permeated with the arrogance of humanism that led us to our current predicament.[1] Instead, all knowledge at whatever scale is reflexively regarded as good even if we do not understand the consequences of its manifestation in the world, have no plan to repair any resulting unanticipated damage, and have no way to hold anyone accountable for damage at scales too large to be repaired. So we confidently rush on where Angels would fear to go and call it progress.

A second, and I think inescapable, reality is that the larger institutional structures of education—schools, colleges, universities, research institutions, and professional organizations—have grown according to the logic of total human mastery and, as Francis Bacon once put it, "the affecting of all things possible." The results are surely mixed. But on the whole, it's gotten us into a heap of trouble. In turn, this logic—the DNA of the system —implies toleration only for incremental changes at the margin as long as they do not threaten the existing structures of power and reputation. Again, not everywhere and always, but all too often. Further, it means that the system of rewards, incentives, promotion, hiring, firing, and funding is rooted in small questions and accordingly averse to large ones of the "emperor is naked" sort. What, for example, is the logic of creating smart and perhaps lethal robots in an overpopulated and underemployed world? Not much, but woe to the young professor of computer science who dares to ask the question or mentions the risks of what philosopher Nick Bostrom calls "super-intelligence." An even worse fate befalls the untenured economics professor who challenges the religion of endless economic growth in a "full world," or the need for redistribution of wealth when some 62 people have more wealth than the bottom 3.5 billion and some of whom lavishly fund institutions of higher learning. And so forth.

A third reality is that time is running out on the experiment of civilization. Climate change and the extinction of species are the surest self-portrait of industrial civilization. There is, in short, no way to read the vital signs of Earth systems with much optimism. To the contrary, they are reasons for the kind of firestorm urgency that should cause a rational species and managers of truly rational institutions to reconsider assumptions, paradigms, laws, regulations, and not the least, its manner of education and act accordingly. The deterioration of forests, waters, wildlife, and soils, however, is a symptom of deeper fault lines in our thinking and they are traceable in large measure to our manner of education that places it bets mostly on more of the same.

This leads to a fourth reality, the date of which I will arbitrarily assign to the opening of the World's Fair in Chicago on May 27, 1933, at or near 9 am, Central Time. Until that time the world had changed at a slow, almost metronomic pace. The world's Fair symbolized the great acceleration under the banner, "Science explores, technology executes, Man conforms." Until recently, education was largely confined to equipping the rising generation with the skills, knowledge, and cultural information suited to that particular culture at a particular point of time. This is not to say that nothing changed. But social, economic, and technological change occurred at a pace that could be understood, more or less, while nature changed hardly at all. The seasons came and went much as before and the human drama played out with relatively slow change in our tools, weapons, methods, goals, and with no discernible change in our fallibilities. So, a Roman farmer in, say, the time of Augustus, would have recognized his counterparts farming nearly 2000 years later. A Roman legionnaire could easily have adapted to military life as a soldier in Napoleon's wars. That is no longer the case. Our science and technology have changed beyond recognition and the Earth is rapidly shifting from the Holocene to something being called the Anthropocene. Our descendants, assuming they exist, will live on what Bill McKibben calls "Eaarth," a more capricious and threadbare planet. Coastlines will shift dramatically. Ecologies will change, forests will disappear, and species will die off at a faster rate. Storms will be larger and droughts longer and more severe. Our goal can no longer be to fashion "the world we want," but rather to cap off the worst that could happen to ensure that there is a habitable world at all. And even that will require a great deal of luck. I do not believe that there is a plausible way around the science underlying that conclusion.[2] Since carbon remains in the atmosphere for periods measured in centuries or millennia, the word "solution" as customarily used is not very useful in discussing climate change. In such rapidly changing ecological, technological, economic, and social changes, what is worth knowing? What is of enduring value? How do we teach?

This leads to a final conclusion. In conditions of climate chaos, morale of both students and teachers will be fragile and exposed to continual erosion from the turbulent, clashing, and cross-currents of our time. How we help our students avoid hopelessness and nihilism? How do we keep our wits about ourselves and sustain our own morale and sanity in such times?

Teachers and educators of all kinds have never faced more daunting challenges and perhaps greater opportunities to bring about systemic changes. The reasons for the former are sketched above; the reasons for the latter are inherent in our capacity for creativity, compassion, and foresight. I am sceptical about the drift of recent technology, but it is possible that properly used, some of it would enable us to create bonds and actions that amplify our capacities to foster positive changes. It could also do exactly the opposite. The difference between these two poles will be decided by how well and systemically we think and what we think about and so will depend very much on education. Without exaggeration it will come down to whether students come through their formal schooling as more clever vandals of the Earth and of each other or as loving, caring, compassionate, and competent healers, restorers, builders, and midwives to a decent, durable, and beautiful future. If the latter, their education must begin in values that stress our connectedness in the fullest sense of the word. And it must enlarge their capacity for affection also in the fullest sense of the word.[3]

In the pages later, some of the most imaginative, skilled, and dedicated educators of our time describe the transition to a new "post-sustainability and environmental education" calibrated to the world of the twenty-first century. In various ways their common aim is to educate a generation of students who grow to be dangerous to the status quo, to injustice in its many forms, to violence visited on humankind and nature alike, to complacency, and to the muddled thinking that conceals the evil men do. Their common aim, expressed in various ways, is to build a far better world that begins in clarity of mind, compassion, dedication, and the stamina to endure undergirded by the awareness "that wonder is, now more than ever, an essential survival skill."[4]

Oberlin, USA   David W. Orr
August 2016

Notes

1. The phrase is taken from the title of David Ehrenfeld's (1980) classic book, *The arrogance of humanism*.
2. See, for example, David W. Orr (2016) *Dangerous years: Climate change, the long emergency, and the way forward*.
3. Orr, 2016, pp. 99–115.
4. See Robert Macfarlane, *Landmarks*, 2016, p. 238.

References

Ehrenfeld, D. (1980). *The arrogance of humanism*. New York: Oxford University Press.

Macfarlane, R. (2016). *Landmarks*. London: Penguin books.

Orr, D. (2016). *Dangerous years: Climate change, the long emergency, and the way forward*. New Haven: Yale University Press.

David W. Orr has been a long-time inspiration for environmental educators. His first two books, *Ecological Literacy* (SUNY, 1992) and *Earth in Mind* (Island Press, 1994), were about education premised on the idea that "all education is environmental education—what is included or excluded teaches that we are part of or apart from the natural world." Since 1990, he has been a Professor in the Environmental Studies Program at Oberlin College, Ohio. In this role, he was instrumental in the design, funding, and building of the Adam Joseph Lewis Center, the first substantially green building on a US college campus and powered entirely by sunlight. David Orr reads for pleasure stimulated by an unruly curiosity, and he writes "to help organize my thoughts and to make what sense I can of the world of the long emergency." His most recent book is *Dangerous Years: Climate Change, the Long Emergency, and the Way Forward* (Yale University Press, 2016). He is currently the Paul Sears Distinguished Professor of Environmental Studies, Emeritus, Oberlin College.

PREFACE

The idea for this book has been incubating for 2 or 3 years. Its importance was underscored at the 2015 World Congress on Environmental Education where I chaired a session titled "Post-Sustainability: Remaking education." A persistent sub-theme from that session was a view that concepts like environmental and sustainability education and education for sustainable development are, in the words of one participant, "debilitated by a lack of philosophical clarity." That participant was Stephen Sterling, my co-editor for this book. Central to his concern, and that of other presenters, is that the task at hand is not to add new bits to the curriculum, or to invent new adjective-driven educations, but rather, to frame a "vision for education" aligned to our extraordinary times.

As the United Nations Decade of Education for Sustainable Development has recently concluded, we now have a new opportunity to reconsider future educational aims. This coincides with a rapid expansion in ecological and environmental interest across the field of education—including curriculum studies and philosophy. And, there is a parallel interest in the public arena. This growing interest is, however, more than just timely. It is urgent, as David Orr so eloquently points out in his Foreword to this book. As he says, "time is running out on the experiment of civilization." Ecologies are changing, forests are disappearing, species are dying, people are dispossessed, and climate is changing.

This book aims to provide serious critiques of education as a whole, and environmental and sustainability education in particular, and then to begin to revision, and indeed remake, a more complete version of education that responds to the challenges of our times. To meet this aim, this volume gathers some of the most prominent international scholars who have been working on these issues over the last three decades.

We believe that this short book will be useful in the most obvious places—courses in environmental and sustainability education. However, the substance of this book is increasingly important at the centre of education—in curriculum studies, educational foundations, and philosophy of education. It might, thus, serve as an introductory reader for remaking twenty-first-century education.

Education also takes place at home, at work, and in community activities—with our children, our peers, our friends, and our neighbours. Education takes place in museums, aquariums, parks, playgrounds, summer camps, and social service agencies. And, of course, it takes place in schools, colleges, and universities. There are educational steps that can be taken by parents, students, community educators, and teachers. There are also steps that can be taken by school principals, curriculum specialists, superintendents, academics, university presidents, education ministers, generals, admirals, presidents, and prime ministers. The time for this collective education action is now. Never before has it been more critical to thoughtfully examine human activities on the Earth—our deepest assumptions, ideals, values, and worldviews. This is work for everyone.

Finally, this is an optimistic book. This is no time for cynics who are more content with despair than hope. As David Orr has so often said, "Hope is not the same thing as wishful thinking. Rather, it is located between wishful thinking and despair." And, "It is a verb with its sleeves rolled up." With hope, hard work, and fresh insights, educators can help to build a better world. Thank you for your engagement, and good luck.

Thunder Bay, Canada   Bob Jickling
October 2016

Contents

Post-Sustainability and Environmental Education: Framing Issues

Bob Jickling and Stephen Sterling

Abstract This book provides a critique of more than two decades of sustained effort to infuse educational systems with education for sustainable development, sustainability education, and, for longer still, environmental education. Additionally, taking to heart the idea that deconstruction is a prelude to reconstruction, this critique leads to discussion about how education can be remade in ways that are conceptually strong and respond to the educational imperatives of our time, particularly as they relate to ecological crises and human/nature relationships. Central to the task at hand is not to add new bits to the curriculum, or new signifiers, but rather, to frame a "new vision for education."

Keywords Environment · Sustainability · Sustainable development · Education · Deconstruction · Reconstruction · Ecological crisis

B. Jickling (✉)
Lakehead University, Thunder Bay, Ontario, Canada
e-mail: bob.jickling@lakeheadu.ca

S. Sterling
Centre for Sustainable Futures (CSF), University of Plymouth, Plymouth, UK
e-mail: stephen.sterling@plymouth.ac.uk

© The Author(s) 2017
B. Jickling, S. Sterling (eds.), *Post-Sustainability and Environmental Education*, Palgrave Studies in Education and the Environment,
DOI 10.1007/978-3-319-51322-5_1

1

"Sustainable excitement," now there's an idea, but what on earth does it mean? Could it be: The normalizing of hyperactivity? A nightmare for teachers? Or, maybe something to give the person with everything? The more the idea is explored, the more ludicrous it seems. Yet, this catchy word play was used with serious intent in an advertisement run in Canada's leading national newspaper.[1] Here it was used to sell Volkswagen's Jetta, the "turbocharged hybrid." Philosopher Hebert Marcuse (1964) calls this the music of salesmanship.

Of course this is not the first provocative use of the word "sustainability." We have also witnessed terms like sustainable economic growth, sustainable mining, sustainable tourism, sustainable consumerism, and even sustainable overfishing. In fact, a study conducted in the years following the Brundtland Commission's *Our Common Future*, published in 1987, found that within 10 years more than 300 definitions for sustainability and sustainable development had already been identified (Dobson, 1996), and a majority of them had been constructed by those with economic interests. The Volkswagen advertisement suggests that the term is as plastic today as it was more than 25 years ago. Most worrisome, however, this plasticity can reinforce and sustain values and attitudes that are harbingers of ecological catastrophe.[2]

This book provides a critique of more than two decades of sustained effort to infuse educational systems with education for sustainable development, sustainability education,[3] and environmental education; because, whatever has been achieved in this period, it is not sufficient. Additionally, taking to heart the idea that deconstruction is a prelude to reconstruction, this critique leads to discussion about how education can be remade in ways that are conceptually strong and respond to the educational imperatives of our time, particularly as they relate to ecological crises and human/nature relationships. Central to the task at hand is not to add new bits to the curriculum, or new signifiers, but rather to frame a "new vision for education."

SUSTAINABILITY AND EDUCATION

To be fair, many educators like to use the term "sustainability." Or, they are at least prepared to go along with it for the time being. For some, sustainability stands in opposition to unsustainability or the relentless destruction of our ecology, especially when ecology is taken as our *oikos*, or household. And, it can stand in opposition to unrestrained free-market capitalism and its cousins, consumerism, consumption, and exploitation. Used in these ways, sustainability can serve as a vantage, or a non-conforming resting place, for critically appraising the status quo.

So, what gives with this Volkswagen advertisement? This kind of representation does make many sustainability educators uneasy. Many of them do recognize plasticity in the term that can render it toothless. After all, sustainability simply means to keep going continuously with little, or no, inherent guidance about what should be sustained.

Some educators have valiantly argued that there is merit in this plasticity. It creates uncertainty and, so goes the argument, this uncertainty represents a dissensus that demands critique, reflection, and discussion—all pathways to new understanding. Still, if this strategy of seeing virtue within dissensus is useful, a big question remains: Is sustainability the most useful term—the term that stands to create the most effective critical traction—to serve educational interests? Or: Will sustainability be sufficient, educationally?

THE PROBLEMATIC OF ENVIRONMENTAL EDUCATION

The history of environmental education, and its predecessors, has always been problematic. Trends within this field have been criticized as behaviourist, excessively consumed with problem solving, sweltering under a thick blanket of scientism, or flirting with instrumentalist ideologies, especially when implementing concepts like education for sustainable development, or "ESD" as it popularly known. For some educators, a response to problems like these has been to break from the constraints and shortcomings of earlier framings.

Today, nature education and conservation education have been to some extent subsumed by environmental education. And, environmental education, itself, has been challenged. Some have distanced themselves from dominant behaviourist tendencies, and dislike of the word "environment," to embrace ecological education. Others believing that environmental education had been constructed too narrowly—especially with regard to social issues—turned to education for sustainable development. Taking another task, yet others have been drawn to ideas about place-based education.

Still other educators seem to sense problems with education for sustainable development and appear to be tilting towards education for sustainability, or sustainability education.

If misgivings about sustainability continue to percolate—prodded by the Volkswagen advertisement, and elsewhere—it, too, may seem insufficient. If history to be is our guide, it is not difficult to imagine the creation of new names—or neologisms—to guide future work. But, educators might ask, is this the most effective path? This book examines that question.

Empty Signifiers

Edgar González-Gaudiano (2005, 2006) introduced the idea of "empty signifiers" in critiquing the rise of education for sustainable development. For him, empty signifiers operate like myths, with pretentions of being salvation narratives. In the end, he argues, empty signifiers are hollow places with little or no inherent meaning. Attempts to fill them with meaning will always be transitory and subject to questioning.[4] He questions this tendency to offer new constructs that will overcome the imperfections and deficiencies of the past solution. This leads him to ask, "is it necessary to create a neologism such as ESD to overcome deficiencies and inertia, or is it a case of the same worn out manoeuver of proposing neologisms which...causes processes of 'change so that nothing changes'"(2005, p. 248)?

Are we just going to continue replacing one empty signifier with another? Is there an alternative strategy? These, too, are questions taken up by this book.

Post-Sustainability and the Remaking of Education

The prefixes "post," "post-post," and "neo-post" represent a recent trajectory in academia[5] in what can be seen, at times, like a manic quest to be new and innovative—to be the first. To those outside of the narrow band of academia, this can all seem confusing, or worse. John Ralston Saul (1995) argues that academic disciplines, such as philosophy, have absented themselves from public debate on important issues. The reason he suggests is that many academics "are caught up in the complexities of philosophical professionalism—a world of narrow specializations and impenetrable dialect. A corporation of philosophy" (p. 161–162). The result, he suggests, is that they have left public debate wide open and vulnerable to more cynical forces.

Here we are hoping to bend the trend in a different, and more user-friendly way. In talking about post-sustainability, we are mainly asking an everyday question: What should we do next? Should we seek to construct ever more-clever signifiers to carry us along for a few more years? And what do we do with the "already many" signifiers such as peace education, environmental education, gender education, human rights education, democracy education, biodiversity education, and sustainability education (amongst others), all of which are trying to "signify"—either implicitly or explicitly—the need for a rethink of the purpose, culture, content, and direction of education. Users of these signifiers often seem to get "caught up" in their own identity deliberations and internal ideological or sociological and sociocultural struggles as they seek to "make an impact" on mainstream education.

It is no surprise that there are numerous parallel efforts being made to "mainstream" all of these signifiers into our education and training systems locally, nationally, and/or globally. Does this suggest a serious misalignment between mainstream education and the cultural, social, and social-ecological issues of the day? Or suggest a new "genre" of institutional politics in the field of education, or both? We have, for example, seen some educators somewhat dogmatically claiming that one prescriptive form of education, namely education for sustainable development, could or should encompass all others? And, some of them, in fits of incestuous naiveté, have asserted that is it is a waste of time to take up the merits of other signifiers.

Similarly, United Nations programmes have seemingly wandered in their own search for meaning and grounding between several emphases. Today, UNESCO has a section called "Education for Sustainable Development and Global Citizenship" which sits under the Division for Inclusion, Peace and Sustainable Development. Formerly, the section was just entitled "Education for Sustainable Development."

More recently, in 2015, the United Nations adopted a 2030 Agenda that is built upon 17 sustainable development goals. One of these goals is simply called "Quality Education" and still centres primarily on access to education rather than reflecting the concerns and experience of decades of work around education, environment, and sustainability—whilst the other 16 goals and their associated targets make very little reference to the role of education and learning. So what is going on? Can we look at fundamental educational questions in another way?

In the subtitle of this book, we suggest talking about how we might remake education for the future. In doing so, we think that endless pursuit of new signifiers will be dissatisfying and ultimately empty. As can be seen in the discussion above, they do important work, but are creating a nest of institutional politics and social movement activity that may not be achieving the depth and breadth of results intended, and necessary for the critical times we have collectively made for ourselves.

Perhaps, it is time to get off the signifier bandwagon and do some more fundamental rethinking of education and its purposes in a rapidly changing global context. This, after all seems to be what the signifiers, including environmental and sustainability education and education for sustainable development have been pointing towards over many years. We could also be asking—is there something fundamentally wrong with education?—how it is perceived, conceived or ignored, constructed, and practised. Does all this talk about sustainability, environmental, and place-based education really tell us that there is something at the core of education itself that is amiss? If so, how would we remake education? In the end, these questions are what this book is about. We would not presume that this book elaborates thorough answers, but it mounts critiques and indicates territory that is promising, significant, and surely worth further exploration in the urgent pursuit of a remade education, appropriate to the signs and needs of our times.

THE CHAPTERS

This book falls into four consecutive sections, the first being "Remaking Education." Bob Jickling's chapter develops themes outlined in this introduction, suggesting that revisiting environmental and sustainability education invites critical reflection on the adequacy of education as a whole to respond to our times, and on whether it can sustain new ways of being in the world. The growth of environmental and sustainability education and allied educational movements for change signal a root problem with dominant educational trends that fail to address contemporary issues. A dissensus is important but is not enough in the search for a "new vision" for education. Rather, Bob Jickling underlines the importance of fostering alternative ways of understanding, and of transformative moments in educational experience. Taking three examples of such moments from literature, he weaves traits and "waypoints" that indicate how transformative education might be taken forward.

Stephen Sterling takes up the theme of remaking education by critiquing the instrumentalist view of the purpose of education manifested in recent decades through the influence of neoliberal ideology. He suggests that education has become maladaptive to the global systemic issues that are shaping the near human and planetary future. A deep learning response within educational thinking, policymaking, and practice is required based upon an emerging relational or ecological worldview, already burgeoning in diverse civil society movements, and indicated by environmental and sustainability education. This would allow attention to be brought to generating purposes and assumptions in education aligned to the possibilities of societal breakdown or breakthrough in global and local systems as the century plays out.

The second section, "Critique and proposition," begins with a philosophical chapter by Sean Blenkinsop and Marcus Morse. These authors draw provocatively on the existentialist philosopher Camus who posits that in choosing not to exercise the radical freedom to commit suicide, people are, in fact, saying yes to life. For Camus, this is a rebellious act that reflects both exaltation of life and negation of death. And, it is the twin act of exaltation and negation that distinguishes rebellion from the more ephemeral results of revolution. The chapter concludes with an exploration of implications for environmental educators who want to consider what it might look like to be a creative, rebel teacher.

Heila Lotz-Sisitka's chapter replies to a recent invitation by UNESCO to respond to their book on the purpose of education, entitled *Rethinking Education: Towards a global common good?* (UNESCO, 2015). The concept of the common good (as it relates to concepts of commons and commoning activity), and what it might mean to engage with commoning as an educational activity, is explored. Drawing on critical realism and decolonization theory, as well as experience of working with expansive social learning, the author proposes that an educational theory grounded in a concept of emergence is needed. Educators are invited to engage critically and imaginatively with the intersection between sustainability, the global common good, and the humanistic orientation put forward by UNESCO in their 2015 document.

The third section "Experience and Relation" centres on the primacy of relationship as central to rethinking education. In his chapter, Michael Bonnett argues that, while the idea of sustainable development fails to provide an adequate way forward for addressing our current environmental crisis, the idea of sustainability contains the germ of an understanding of the character of human consciousness that places our relationship with nature at the heart of both human being and authentic education. A phenomenological approach is developed to explore some key aspects of our experience of nature—particularly its "otherness" and intrinsic normativity. The potential of the mutually sustaining relationship with nature that then emerges, including countering the influence of scientism, is discussed along with broad implications for a re-envisioned education.

The second chapter in this section is by Lesley Le Grange, who analyses and discusses the possible meanings of the idea "after sustainability." The first meaning—in pursuit of sustainability—is explored by regarding sustainability in metaphor as a rhizome, as an empty signifier, and as the potentia of sustainability culture (a grass-roots societal movement). The author proposes that this interpretation offers radical alternatives to dominant discourses on sustainability. Second, the author reflects on the possibility of moving beyond the idea of sustainability, informed by an ontology of immanence whereby the subject becomes imperceptible, and so too does the idea of sustainability.

The book finishes with a flourish in the final "Education through Action" section, where the chapter authors ground their thoughts on remaking education in the real-world struggles of communities. Lucie Sauvé's insightful chapter stems from her experience with citizen resistance movements against invasive projects, and her exploration of diverse ecosocial creative initiatives. Here, a rooted learning dynamic and a collective intelligence have emerged that stand in contrast to the experience of most schooling in the formal sector. The author argues that formal education could benefit from non-formal and informal education so as to empower youths/citizens willing and capable of contributing to social debates and transformations for better living together within our *oikos*, our common home.

In the final chapter, Edgar J. González-Gaudiano and José Gutiérrez-Pérez denounce the effects of extractive megaprojects (opencast mining, microdams, shale gas extraction), including several preconized as alternative production and sustainable energy strategies (giant wind turbines), resulting in social conflicts as a result of the breakdown of community ties, the destruction of regional economies, the loss of cultural diversity, and the degradation of environment. In this context, the familiar challenges faced by environmental education are magnified and assume an urgency and immediacy. In response, the authors advocate "resilient education" as a socio-critical, emancipatory, and political approach focussing on the skills that will be increasingly required to tackle environmental degradation and climate change taking place in the world, especially in the so-called developing world through, citizen engagement.

In "An Afterword," Jickling and Sterling reflect on some of the main themes emerging from the collected chapters. The UNESCO Global Monitoring Report (UNESCO, 2016), which was launched just as the first draft of this book was completed, is welcomed as it couples the status of education with planetary prospects. Yet, it also serves to underscore the very purpose of this book by failing to recognize the nature and depth of change required in educational practice to meet the aspirations of the report's subtitle, "creating sustainable futures for all." Rather, as reflected by the different authors in this book, there needs to be a disruption of dominant assumptions in educational thinking and purpose so that a cultural shift towards practice that is life-affirming, relational, and truly transformational can take root. This can be realized at any level of engagement through the role of "rebel teacher" and through "being differently" in the world.

An Agent in Discovery

It might be tempting to think of this book as a framework for the future, but that would not be correct. We hope, rather, that it will be seen as a heuristic. These two terms are different in important ways. Frameworks provide more concrete visions about how things are, how they should be, or roadmaps for getting to a new place. But heuristics are agents in the process of discovery—provocateurs at the intersection of imagination and praxis.

In truth, as we move between geological epochs—between the Holocene and the Anthropocene—we are traversing new terrain. Humans have never before witnessed this kind of epochal shift, or had to accept this scale of responsibility. No one knows what will happen or how we will need to respond; uncertainty is defining. We do hope that the heuristic agency of this collection will inspire responses that are imaginative, creative, courageous, and radical—because this is what our times require.

Notes

1. The Glove and Mail, April 1, 2013, p. A14.
2. See, for example, Blühdorn, 2002, 2011 and Jickling, 2013.
3. Here we are referring to Initiatives arising out of the Rio Earth Summit and Agenda 21 in 1992 and peaking with the decade of education for sustainable development that has just ended.

4. There has been a recent defence of empty signifiers and education for sustainable development. See Bengtsson & Östman, 2013 and Bengtsson, 2016. This defence will be examined in later chapters of this book.
5. See, for example, Patti Lather, 2013.

REFERENCES

Bengtsson, S. L. (2016). Hegemony and the politics of policy-making for education for sustainable development: A case study of Vietnam. *Journal of Environmental Education, 47*(2), 77–90.

Bengtsson, S. L., & Östman, L. O. (2013). Globalisation and education for sustainable development: Emancipation from context and meaning. *Environmental Education Research, 19*(4), 477–498.

Blühdorn, I. (2002). Unsustainability as a frame of mind: And how we disguise it: The silent counter-revolution and the politics of simulation. *The Trumpeter, 18* (1), 59–69.

Blühdorn, I. (2011). The politics of unsustainability: COP15, post-ecologism, and the ecological paradox. *Organization & Environment, 24*(1), 34–53.

Dobson, A. (1996). Environment sustainabilities: An analysis and a typology. *Environmental Politics, 5*(3), 401–428.

González-Gaudiano, E. J. (2005). Education for sustainable development. *Policy Futures in Education, 3*(3), 243–250.

González-Gaudiano, E. J. (2006). Environmental education: A field in tension or in transition? *Environmental Education Research, 12*(3–4), 291–300.

Jickling, B. (2013). Normalizing catastrophe: An educational response. *Environmental Education Research, 19*(2), 161–176.

Lather, P. (2013). Methodology-21: What do we do in the afterword? *International Journal of Qualitative Studies in Education, 26*(6), 634–645.

Marcuse, H. (1964). *One-dimensional man.* Boston: Beacon Press.

Saul, J. R. (1995). *The unconscious civilization.* Concord, ON: Anasi.

UNESCO. (2015). *Rethinking education: Towards a global common good?* Paris: UNESCO. Retrieved from http://unesdoc.unesco.org/images/0023/002325/232555e.pdf

UNESCO. (2016). *Education for people and planet: Creating sustainable futures for all.* Paris: UNESCO. Retrieved from http://unesdoc.unesco.org/images/0024/002457/245745e.pdf

Bob Jickling has been an active practitioner, teaching courses in environmental philosophy; environmental, experiential, and outdoor education; and philosophy of education. He worked as Professor of Education at Lakehead University after many years of teaching at Yukon College. He continues his work as Professor Emeritus at

Lakehead University. His research interests have long included critiques of environmental education and education for sustainable development while advancing formulations for opening the process of remaking education.Jickling was the founding editor of the *Canadian Journal of Environmental Education* in 1996, and together with Lucie Sauvé, he co-chaired the fifth World Environmental Education Congress in Montreal, in 2009. He has also received the North American Association of Environmental Education's Awards for Outstanding Contributions to: Research (2009) and Global Environmental Education (2001). In 2012, he received the Queen Elizabeth II Diamond Jubilee Medal in recognition of contributions to Canada. As a long-time wilderness traveller, much of his inspiration is derived from the landscape of his home in Canada's Yukon.

Stephen Sterling is Professor of Sustainability Education, Centre for Sustainable Futures (CSF) at the University of Plymouth, UK. His research interests lie in the interrelationships between ecological thinking, systemic change, and learning at individual and institutional scales to help meet the challenge of accelerating the educational response to the sustainability agenda. He has argued for some years that the global crises of sustainability require a matching response from the educational community, together with a shift of culture towards educational policy and practice that is holistic, humanistic, and ecological. A former Senior Advisor to the UK Higher Education Academy on Education for Sustainable Development (ESD) and current National Teaching Fellow (NTF), he has worked in environmental and sustainability education in the academic and NGO fields nationally and internationally for over three decades. He was a member of the UNESCO International Reference Group for the UN Decade on Education for Sustainable Development (2005–2014), and he is currently co-chair of the International Jury for the UNESCO-Japan Prize on ESD. Widely published, his first book (co-edited with John Huckle) was *Education for Sustainability* (Earthscan, 1996), and this was followed by the influential Schumacher Briefing *Sustainable Education—Re-visioning Learning and Change* (Green Books 2001).He also co-founded the first masters course in the UK on sustainability education (at London South Bank University), and led the WWF-UK project on systems thinking *Linking Thinking—new perspectives on thinking and learning for sustainability.*

His work at Plymouth involves developing strategies to support whole institutional change towards sustainability. Other books include (with David Selby and Paula Jones Earthscan, 2010) *Sustainability Education: Perspectives and Practice across Higher Education* and (with Larch Maxey and Heather Luna) *The Sustainable University—Process and Prospects,* published by Routledge in 2013.

PART I

Remaking Education

Education Revisited: Creating Educational Experiences That Are Held, Felt, and Disruptive

Bob Jickling

Abstract First, broad issues in education are discussed and similarities with issues in environmental and sustainability education are noted. Second, some key initiatives, arising after the conclusion of the Decade of Education for Sustainable Development, are noted and discussed. These include the United Nation's 2030 Agenda and its implications for education, and a special issue of *Journal of Environmental Education* devoted to a critical examination of the politics of environmental education and education for sustainable development. Analysis suggests that environmental education and sustainability education can both be seen as signs that point to the question, "What is fundamentally wrong with education itself?" Third, a thought experiment is presented to tackle this question. Quotations from Arne Næss, Aldo Leopold, and Albert Camus are used to probe transformative moments in these authors' lives. They are found to point towards something that is undervalued in much of contemporary education.

B. Jickling (✉)
Lakehead University, Thunder Bay, Ontario, Canada
e-mail: bob.jickling@lakeheadu.ca

© The Author(s) 2017   15
B. Jickling, S. Sterling (eds.), *Post-Sustainability and Environmental Education*, Palgrave Studies in Education and the Environment,
DOI 10.1007/978-3-319-51322-5_2

Keywords Environment · Sustainability · Sustainable development · Education · Transformation · Experiential · Feeling · Emotion · Resonant understanding

As I began to put my emerging thoughts to paper, I was evermore convinced that this book is timely. Evidence is all around. Just a day earlier, I opened my most recent issue of *Educational Researcher* and read the lead article by Kris Gutiérrez (2016). She is a former President of the American Educational Research Association, and in this paper she wrote that what concerns her most is:

> our inability to intervene productively, at least in any sustained and trans-
> formative way, in the academic lives of so many youth today—to imagine
> new trajectories and futures and forms of agency.... (p. 187)

Here, she was particularly concerned about youth in vulnerable circumstances. But, her concerns are more broadly applicable. Gutiérrez goes on to say:

> we simply cannot rely on efficiency and market-driven models of education
> that are certain to bankrupt the future of our nation's youth. We need
> models for educational intervention that are consequential—new systems
> that demand radical shifts in our views of learning.... (p. 187)

I begin with Gutiérrez to highlight broad concerns about education. For her, imagining new trajectories for interpreting and acting in the world are part of an educational vision. And, learners acting with new agency reflect experiences of transformation. I also quote Gutiérrez in order to illustrate how similar her concerns are to those of many environmental and sustainability educators.

Like many environmental and sustainability educators, Gutiérrez also sees current models of education to be part of the problem. In addition to her work, criticisms of market-driven models are enjoying fresh attention elsewhere. Wayne Au's widely read paper (2011) lays bare the effects of "the 'New Taylorism', where [teachers'] labour is controlled vis-à-vis high-stakes testing and pre-packaged, corporate curricula aimed specifically at teaching to the tests" (p. 25). His research shows that in the United States, high-stakes testing "not only standardizes the content of the curriculum as well as the forms such content takes in the classroom, it also works to standardize teachers' pedagogies as they work to deliver test-driven curriculum in an efficient manner" (p. 31). Parallel research suggests that these trends are not isolated, but part of a "global testing culture" (e.g. Smith, 2016).

Criticisms of market-driven, outcomes-oriented, factory production, industrial, or efficiency curricula provide clues to systemic barriers to transformative educational experiences. However, I see citizen movements pushing back against standardized testing. Au reports that teachers, too, "often create space for what they consider to be 'real teaching' in the face of the high-stakes testing pressures" (p. 39). Gutiérrez's own social design experiments, discussed later in this chapter, also offer alternative examples with transformative potential.

In revisiting environmental and sustainability education we are revisiting concerns about visions of education itself. For example, what would it take to educate in a sustained and transformative way? And what are some of the barriers to implementing this kind of education?

Transformation is a large and controversial topic—too large for a single chapter. So here, I limit my use of "transformative" to the kinds of educational experiences that allow learners to sustain new ways of being in the world. First, however, I examine briefly some key events in environmental and sustainability education.

Issues in Environmental and Sustainability Education

Following the end of the United Nations Decade of Education for Sustainable Development, in 2014, there is a small political vacuum that offers a window for reflection and imagination about environmental and sustainability education.

In this section, I sketch some key initiatives of this post-decade period as context for this chapter and for insights into paths forward. In particular, I am interested in key United Nations-led developments and a reinvigorated international debate over the relative merits of environmental education and education for sustainable development.

The United Nations

In September 2015, the United Nations approved its 2030 Agenda with an ambitious aim of "transforming the world." A road map to this aim is laid out in 17 sustainable development goals. Most of these goals point to important areas of social and environmental health, and justice. Goal 4 calls for Quality Education.

The 2030 agenda was hailed by the Secretary-General, Ban Ki-moon, as a "promise by leaders to all people everywhere. It is an agenda for people to end poverty in all its forms—an agenda for the planet, our common home."[1] This international agreement identifies many important priorities—like poverty, hunger, inequalities, climate change, and clean energy, to name a few. These goals provide educationally worthwhile areas of concern. But, implicit in the nature of the United Nations Resolution is compromise. An important educational question remains: What is missing?

The fourth sustainable development goal is to "Ensure inclusive and quality education for all, and to promote lifelong learning" (United Nations, 2015, p. 14). Most of the targets for this goal are aimed at access to education and equity. However, one target aims to ensure that all learners acquire the knowledge and skills needed to promote sustainable development. There is no conceptual framing of quality education, and no identification of the required knowledge and skills. But this language does hint at a standardized, outcomes-driven approach to curricula.

Indeed, this outcomes-driven approach was prominent at the first educational conference after the launch of Agenda 2030, hosted in January 2016 at the Centre for Environmental Education in Ahmedabad, India. There was much talk at this conference about "Education as a Driver for Sustainable Development Goals." Also prominent were discussions about how progress would be indicated and measured. Interestingly, an older gentleman—an elder—stood up and responded to a brutally dull session on measurement indicators and teaching progress. He talked about the importance of caring, feelings, and compassion. He also hoped that we would not measure ourselves into oblivion and indicate ourselves into insomnia. Insightful.

Unfortunately, the conversation about implementation of quality education pivots towards a version of market-driven education. Identifying "indicators of success" in a model where education is simply a driver for implementing the sustainable development goals is not a radical shift in collective views about learning and education. In this scenario, measuring ourselves into oblivion is a more likely outcome than significant transformation.

In May 2016, Mr. Anote Tong, the now retired President of the small island state of Kiribati, delivered a moving presentation at the "Symposium on environment and displacement: Root causes and implications,"[2] hosted at the United Nations Environment Assembly. He contended that even if carbon emissions are reduced to zero, now, the oceans will continue to rise, and at the end of this century his whole country will be submerged. Prudently, representatives of Kiribati have been looking for places to relocate. Sadly, no one wanted them. Finally, Fiji extended a humanitarian offer, promising a place for Kiribatians.

What this tragic story tells us is that at international levels, nation states—with the exception of Fiji—do not care about the people of Kiribati. By extension, it is sobering to think that in spite of the Sustainable Development Goals, the success of the 2030 agenda will likely fail if nation states and their citizens do not care—about poverty, hunger, justice, equity, or the more-than-human world. At this point the Indian gentleman's point about caring is prescient.

An International Debate

As Guest Editor, Phillip Payne led the *Journal of Environmental Education* in publishing a Special Issue that reflects critically on the politics of environmental education and education for sustainable development (2016a). Payne is concerned about a kind of uncritical mashing together of environmental education and education for sustainable development into what he calls a "marriage of academic convenience" (Payne, 2016b). For him, a critical conversation about the evolving roles of these so-called fields is long overdue. He also suggests that this uncritical mashing of fields is entwined in a stealth revolution of neoliberalism.

This special issue is essentially a debate in print. Stefan Bengtsson (2016a) provides a lead article that explores education for sustainable development within the context of a Vietnamese case study that, in turn, examines the impact of sustainable development policies in this country. Six responses to this lead article are provided, and these are followed by a brief rejoinder by Bengtsson. Payne and his collaborators bring "fresh new steam" into discussions about environmental education and education for sustainable development, and tensions therein. Importantly, this Special Issue also asks what next?

In Bengtsson's (2016a) lead paper, and elsewhere (Bengtsson & Östman, 2013), he responds to critique of education for sustainable development. In particular, he takes issue with González-Gaudiano's (2005) argument that education for sustainable development is yet another empty signifier, devoid of meaning. Rather, Bengtsson puts stock in the presence of contestation and dissensus inherent in the concepts of sustainable development and education for sustainable development. He makes a virtue of the vagueness of the terms and, instead, considers this generative. He contends that meaning will develop in multiple and intersecting ways through diverse discourses in the field. Interesting, but does it actually do this?

There is dissensus—enough to feel good—but are there any signs of transformative potential? Does it offer the kinds of educational experiences that allow learners to sustain new ways of being in the world? Is dissensus— any dissensus—enough? For example, studies about the relative merits of the Madrid and Barcelona football teams would certainly generate dissensus. But, it is important to ask, is this topic, with its inherent dissensus, sufficiently worthwhile to justify inclusion in an educational programme? I would not want to underestimate the ability of scholarly football fans to find ways to make this important, but for the most part it seems a fair question.

Sadly, there is little to celebrate when reviewing Bengtsson's Vietnamese case studies. The representation of the environment and human/nature relationships in policy documents is dismal, as are disparities between rich and poor, attributed to a rapidly increasing market economy. In short, it appears that sustainable development, as an organizing framework and provocateur of dissensus, has had little traction in problematizing the dominant economic discourse. If anything, the evidence provided suggests a stealth victory for neoliberalism. Indeed, Bengtsson acknowledges in his own rejoinder that all participants in the special issue's debate share the view that "sustainable development doesn't seem to have gained much traction in education in the ways its policies and practices aimed for" (2016b, p. 163).[3] He also claims that the same might be said of environmental education.

This debate suggests two important considerations. First, following Gutiérrez's concern at the beginning of this chapter, and then picking up on Bengtsson's critical point, it may be that neither education for sustainable development, nor environmental education, has been able to intervene in sustained and transformative ways. Second, if simple dissensus is not enough, what is missing? This question is just an opening—an invitation to look deeply into what is fundamentally wrong with education as it is most often experienced at this time. It also invites us to pivot away from the long-standing debate between environmental education and education for sustainable development, and come at educational questions from a different direction.

Where Do These Observations Lead?

Finally, there is a reopening of conversations about the nature, purpose, and importance of environmental education. This is happening on multiple fronts, but where will we go with it? Since the early 1980s there have been arguments about whether environmental education has definition and structure. And the evolution of environmental education has been described as a typology of currents and framed as fields. Similar patterns have followed as education for sustainable development has struggled to establish its own legitimacy. We could continue these kinds of processes. But, I am not sure that the outcomes would be useful.

Here, I propose another approach. Where could we go if we thought of sustainability and environmental education less as signifiers of new fields within education, but rather as signs that something is missing—indeed fundamentally wrong—with education itself? Seen this way, environmental and sustainability education are not necessarily competitors for limited educational space, but rather signs that point to the same questions.

WHAT COULD A TRANSFORMATIVE MOMENT LOOK LIKE?

The questions I raise are persistent. To address them, I return to an old observation to see what new light can be shed. While some might say that earlier philosophers of education have been overtaken in the present era, they still left some useful observations. For example, R. S. Peters suggested that it would be unreasonable, "to deprive anyone of access in an arbitrary way to forms of understanding which might throw light on alternatives open to him (sic). This is the basic argument for breadth in education" (1973, p. 256). In our contemporary context, as suggested by the analysis above, it still matters to ask, what is important but has been left out? What new, or different, understandings are required if education has any chance of transforming ways of being in the world?

To tackle these questions, I propose a thought experiment. Here, I present three textual examples from historic thinkers who have all, in one way or another, pondered their own transformations. One quotation is from a work of fiction and the other two are from essays more closely aligned with environmental thinking. I suggest that these vignettes all describe transformative moments.

Arne Næss

Arne Næss was a Norwegian philosopher who, after his formal retirement, devoted himself to eco-philosophy or, as he preferred to call it, ecosophy. Consider his reflections:

> What would be a paradigm situation of identification? It is a situation in which identification elicits intense empathy. My standard example has to do

with a non-human being I met 40 years ago. I looked through an old-fashioned microscope at the dramatic meeting of two drops of different chemicals. A flea jumped from a lemming that was strolling along the table and landed in the middle of the acid chemicals. To save it was impossible. It took many minutes for the flea to die. Its movements were dreadfully expressive. What I felt was, naturally, a painful compassion and empathy. But the empathy was *not* basic. What *was* basic was the process of identification, that "I see myself in the flea." If I was alienated from the flea, not seeing intuitively anything resembling myself, the death struggle would have left me indifferent. So there must be identification in order for there to be compassion and, among humans, solidarity. (Næss, 1988, p. 22)

Næss repeatedly points to this experience as one that has shaped the contours of his thinking—and perhaps we can even say his transformation. In this recognition of suffering he began to see the world differently. He recognized the flea's suffering and his own empathy; and, he continued to work with these revelations for more than four decades.

For Næss, ecosophy is rooted in deeply intimate relationship that shifts one's concept of self from an egotistical "self" to the more expansive "Self" as an expression of identification, relationship, and compassion. The accompanying "Self-realization," as he called it, can be described as an ecological approach to being in the world.

Aldo Leopold

Aldo Leopold was an American conservationist whose work was made famous through his book, *A Sand County Almanac* (Leopold, 1949/1966). An important moment in his life is captured in the following reflection from this book:

Only the mountain has lived long enough to listen objectively to the howl of a wolf.... My own conviction on this score dates from the day I saw a wolf die. We were eating lunch on the high rimrock, at the foot of which a turbulent river elbowed its way. We saw what we thought was a doe fording the torrent, her breast awash in white water. When she climbed the bank toward us and shook out her tail, we realized our error: it was a wolf....

In those days we had never heard of passing up the chance to kill a wolf. In a second we were pumping lead into the pack, but with more excitement than accuracy.... When our rifles were empty the old wolf was down, and a pup was dragging a leg into impassable slide rocks.

We reached the old wolf in time to watch a fierce green fire dying in her eyes. I realized then, and have known ever since, that there was something new to me in those eyes—something known only to her and to the mountain. I was young then and full of trigger-itch; I thought that because fewer wolves meant more deer, that no wolves would mean hunters' paradise. But after seeing the green fire die, I sensed that neither the wolf nor the mountain agreed with such a view. (p. 138–139)

Leopold's early academic influences were shaped by the Yale School of Forestry. This was the programme founded by the "wise use" advocate Gifford Pinchot. For Pinchot, the world was, in short, a resource for human use and our responsibility was to use it wisely. In keeping with this thinking, Leopold's early experiences in forest management included killing predators. Then there was the dying wolf. This moment fell outside of his normal experiences, and he wrestled with it for the rest of his life. As Leopold's thinking evolved, he eschewed ideas that rested on the presumption of human dominance. And he gave us the idea that, "We can be ethical only in relation to something we can see, feel, understand, love, or otherwise have faith in" (1966, p. 251).

Albert Camus

Albert Camus was a North African philosopher and novelist. In his novel, *The Plague* (1947/2013), Camus describes the transformative moment for one of his characters who, as a boy, witnessed his father, a judge, at work:

But, I suddenly became aware of him, though up to then I had only thought of him in the convenient category of the "accused." I cannot say that I forgot about my father, but there was something in the pit of my stomach which distracted me from concentrating on anything except the man in the dock. I heard practically nothing. I felt that they wanted to kill this living man and an instinct as powerful as a tidal wave swept me to his side with a sort of blind obstinacy. I only properly woke up when my father began his speech for the prosecution.

Transformed by his red robe, he was neither good-natured nor affectionate and his mouth was crammed with sonorous phrases which leapt from it constantly like serpents. I realized then that he was asking for the death of this man in the name of society and that he was even demanding that his head be cut off. In truth, all he said was "This head must fall." But the difference was not so great in the end. In fact it came to the same thing, since he got the man's head. The only thing was that he did not do the job himself. I, who subsequently followed the matter through to its conclusion, felt a far more terrifying intimacy with that unfortunate man than my father ever could. And yet he had to be present, according to custom, at what are euphemistically called the last moments, and which one should by rights call the most shameful of murders. (p. 191)

Camus portrays this character as a boy profoundly shaken by an emotional childhood experience. His transformation is affirmed for the reader, years later in his life, through the act of volunteering for the dangerous job of caring for quarantined family members of bubonic plague victims.

A Brief Analysis

First, consider the transformational experiences of these authors. Næss was still talking about his formative experience 40 years after the death of the flea. Leopold's wolf died early in his professional career, yet this passage was published decades later in his final work. And Camus, unsettled by the Algerian War, became a vocal opponent of the death penalty. Each of these authors had significant experiences that fell out-side of their regular norms and expectations. The experiences disrupted deeply held—and previously under-recognized—truths. They were inex-plicable, yet these authors hung onto them; they cogitated about what they meant; they looked to them for meaning; and, they gave them a new perspectival lens through which to view the world anew. In turn, these experiences transformed their lives.

Second, it is not uncommon for readers to feel moved by some, or all, of these passages. But why? Close scrutiny of the text is, I think, dis-appointing. The first quotation from Arne Næss's essay is pretty normal in academia. Taken out of context, "What I felt was, naturally, a painful compassion and empathy," can be seen as a dry and rational recounting. Can we say that we really feel something when we read Camus's, "but there was something in the pit of my stomach which distracted me from concentrating on anything except the man in the dock?" Finally, what should we make of Leopold's line, "But after seeing the green fire die, I sensed that neither the wolf nor the mountain agreed with such a view?" Read literally, Leopold can seem just weird. Who has seen "green fire" in the eye of a wolf and what are we to make of a sentient mountain? But, in some ways he is the most interesting. What I read in this enigmatic passage is a struggle to find words to talk about experiences that fall outside of everyday conventions.

While these authors may seem enigmatic, they have told their stories in ways that resonate with many readers. They point to something already within us. When these stories are placed beside our own stories, we can be suddenly pierced with recognition and understanding—with "that vibrant spoor of what cannot be said" (Lee, 2010, p. 22). Jan Zwicky (2003) calls this a form of ontological appreciation—a resonance that signifies we have experienced a similar way of being in the world, however briefly.

Educationally, this resonance throws light on alternative ways of understanding and being in the world. And, this resonance points to alternative ways of understanding—ways of understanding that allow learners to intervene in their worlds productively, and in a sustained and transformative way. The stories point to something that seems absent and unvalued in much of contemporary education.

Some might complain that the examples are exceptional—composed by celebrated scholars—thus setting the bar too high. I do not think so. Kris Gutiérrez's (2016) own social design experiments point to wonderful pedagogical possibilities when university students and school children work together. In her own words, "They were brought together through an intervention that privileged joint activity, playful imagination, and a vision of teaching in which an imagined or projected future could influence activity in the present" (p. 192). The impact and durability of the experience she recounts suggests transformation, as noted in an affirming email sent to her 13 years later by one of the participants.

Stephanie Kaza (2002) provides another example. She found that her students had a history of concern about racism and other issues of social justice but an ambivalence, or reserve, about activist organizations. Yet, a field trip, labelled a "toxic tour" of environmental justice sites served as an avenue for transformation. Through the direct experience of inequity, they gained an emotional engagement with environmental injustice. Through this experience, the abstract became a real phenomenon affecting real people. The result, to Kaza, was an increase in the level of ethical accountability on her campus implying a subtle transformation in student ways of being.

David Jardine (1998) challenges readers to learn through voices in the more-than-human world. His paper explores the ecological images hidden within—part of his essential being—that were revealed during a bird-watching excursion to the place where he was raised. He gently guides readers to consider teachings revealed through this birding experience, his deeply held intimacy with birds and cicadas and, indeed, his own transformation. More recently, I have been in touch with an Indian educator who recounted a childhood encounter of being caught in the act of killing ants. His grandmother's admonishment, timed while he was still gripping his ant-squashing flip-flop, led ultimately to a life-changing transformation. A direct result was a 20-year commitment to vegetarianism (S. Sugandh, 2016, personal communication).

Similarly, a South African blogger recounted the experience of finding a rhinoceros that had its horn removed by poachers (X. Mali, 2014, personal communication). This, too, points to the possibility of something deeply transformational arising from his encounter. In this case it is probably too early to tell if the event was truly transformational. Often this requires looking back, after a time, to see what personal change transpired. Transformation is hard, often painful, and ultimately visible in hindsight.

When surveying the initial three examples it is possible to trace some entwined traits running through them, and others too. Michael Derby's ecopoetic work, for example, also, provides a delightful example of what it means to keep "knowing connected to the world as it is lived" (2015, p. 20). My provisional list of traits suggests:

- There is something deeply visceral and intimate that runs through these examples.
- They are profoundly sensual and arise out of first-hand experience; they require being in the world. And, they can remind us that we are always and already in the world.
- They evoke emotional engagement, care, empathy, and identification. They can also evoke sadness, disenchantment, and anguish.
- They are relational—ecologically, biotically communal, Self-realizational.
- They point to understandings that are beyond words. These understandings are not descriptive, analytical, logical, falsifiable, or narrowly rational. Yet they exist and are inextricably part of the beings we are.
- While the experiences are beyond words, the stories that we tell about them can evoke resonant understanding. There is a verisimilitude in resonance.
- And, many of these examples reveal listening to, and learning from, a kind of voice from the more-than-human world.[4]

Such a collection of traits has educational implications. If, as Michael Derby suggests, "We have come to experience 'school life' and learning as fundamentally prosaic; characterized by fragmentation, emotionlessness and exacerbated by the privileging of epistemic foundations such as anthropocentrism, reductionism, linear causality, and dualism" (2015, p. 25), then there is a lot of work to do. Consider, for example:

- The understandings pointed to in this collection of traits cannot be abstracted, reduced, or taught in isolation from the world. To gain access learners must be experientially present.
- The outcomes are wild. They defy prediction and control; they just arise from the experience and are suddenly present. We do not create transformative moments, but can create spaces for them to arise (e.g. Gutiérrez, 2016; Jickling, 2016, 2015; Kaza, 2002).
- Given that transformative moments exist beyond the capacity of language to fully describe, it is doubtful that they can be measured or evaluated in a schooling context. Any analysis is possible only in hindsight. Even though this understanding cannot be measured, it still exists. Inclusion will require more than market-driven and outcomes-oriented visions of education.
- Even though the resultant understandings are educationally unmanageable, they can still be educationally transformational. Indeed, the facility to care nurtured in these examples may actually be a prerequisite to transformation.

WHERE DO WE GO FROM HERE?

Au (2011) suggests that in spite of high stakes-testing pressure, teachers often create space for what they consider to be "real teaching." Though, he reports, this usually requires some kind of deception. I have tried to support such committed teachers with another approach. It begins with a question. What is essentially missing in education these days? This question is asked in the face of persistent observations about our inability to sustain anything like transformational educational experiences.

The examples provided are knitted together with some common themes, traits, and considerations. Once identified, they provide waypoints for developing praxis. These waypoints are not strictly logical; many points require shared experience, emotional openness, and resonant understanding. There is no recipe here. It is likely that the types of examples given will not work well in every context. Remaking education will require the hard work of figuring out what is right in your time and place.

As humans, we have the capacity to feel, empathize, love, and mourn loss. When we lose these qualities, we are reduced. When we are reduced this way, things slip by us. We need to pay attention. It might, after all, be feeling that surpasses language, and understanding as we presently know it, that will keep humans from wiping out each other, and all other beings in their way.[5] Thus, creating educational experiences that are held, felt, and disruptive might just be the basis for learning that is, indeed, transformational.

NOTES

1. Downloaded, May 13, 2016: http://www.un.org/apps/news/story.asp? NewsID=51968#.VwH2U84mXCO
2. Downloaded, June 19, 2016: http://web.unep.org/unea/special-events/ symposium-on-environment-and-displacement-root-causes-and-implications
3. Others have made the same observation: for example, Wals, 2009; Huckle & Wals, 2015.
4. I realize that bullet pointed lists are archetypally part of the linear approach to education that this chapter seeks to counter. Imagine then, these points drawn freely as interrelating nodes—like spider webs or rhizomes.
5. Here I am riffing off of Alexis (2015, p. 170), and exploring the value of fiction as research.

REFERENCES

Alexis, A. (2015). *Fifteen dogs.* Toronto: Coach House Books.

Au, W. (2011). Teaching under the new Taylorism: High-stakes testing and the standardization of the 21st century curriculum. *Journal of Curriculum Studies, 43*(1), 25–45.

Bengtsson, S. L. (2016a). Hegemony and the politics of policy-making for education for sustainable development: A case study of Vietnam. *Journal of Environmental Education, 47*(2), 77–90. doi:10.1080/00958964.2015.1021291

Bengtsson, S. L. (2016b). Aporias, politics of ontology, ethics, and "we"? *Journal of Environmental Education, 47*(2), 163–168. doi:10.1080/00958964. 2015.1124830

Bengtsson, S. L., & Östman, L. O. (2013). Globalisation and education for sustainable development: Emancipation from context and meaning. *Environmental Education Research, 19*(4), 477–498.

Camus, A. (2013). *The plague* (R. Buss, Trans.). London: Penguin Books. (Original work published 1947).

Derby, M. D. (2015). *Place, being, resonance: A critical ecohermeneutic approach to education.* New York: Peter Lang.

González-Gaudiano, E. (2005). Education for sustainable development: Configuration and meaning. *Policy Futures in Education, 3*(3), 243–250.

Gutiérrez, K. D. (2016). Designing resilient ecologies: Social design experiments and a new social imagination. *Educational Researcher, 45*(3), 187–196.

Huckle, J., & Wals, A. E. J. (2015). The UN decade of education for sustainable development: Business as usual in the end. *Environmental Education Research, 21*(3), 491–505.

Jardine, D. (1998). Birding lessons and teachings of cicadas. *Canadian Journal of Environmental Education, 3*, 92–99.

Jickling, B. (2015). Self-willed learning: Experiments in wild pedagogy. *Cultural Studies of Science Education, 10*(1), 149–161. doi:10.1007/s11422-014-9587-y

Jickling, B. (2016). Wild pedagogies: A floating colloquium. *Pathways: The Ontario Journal of Outdoor Education, 28*(4), 4–7.

Kaza, S. (2002). Teaching ethics through environmental justice. *Canadian Journal of Environmental Education, 7*(1), 99–109.

Lee, D. (2010). The music of thinking: The structural logic of "Lyric Philosophy." In M. Dickinson & C. Goulet (Eds.), *Lyric ecology: An appreciation of the work of Jan Zwicky* (pp. 19–39). Toronto: Cormorant Books.

Leopold, A. (1966). *A Sand County almanac: With essays on conservation from Round River.* New York: Sierra Club/Ballantine. (First published in 1949/1953).

Næss, A. (1988). Self realization: An ecological approach to being in the world. In J. Seed, J. Macy, P. Fleming, & A. Næss (Eds.), *Thinking like a mountain: Towards a council of all beings* (pp. 19–30). Gabriola Island, BC: New Society Publishers.

Payne, S. L. (Guest Ed.) (2016a). *Journal of Environmental Education, 47*(2).

Payne, S. L. (2016b). The politics of environmental education. Critical inquiry and education for sustainable development. *Journal of Environmental Education, 47*(2), 69–76. doi:10.1080/00958964.2015.1127200

Peters, R. S. (1973). Aims of education: A conceptual inquiry. In R. S. Peters (Ed.), *The philosophy of education* (pp. 11–57). Oxford: Oxford University Press.

Smith, W. C. (Ed.). (2016). *The global testing culture: Shaping educational policy, perceptions, and practice.* Oxford: Symposium Books.

United Nations. (2015). Resolution adopted by the General Assembly on 25 September 2015. *Transforming our world: The 2030 agenda for sustainable development.* Retrieved from http://www.un.org/ga/search/view_doc.asp?symbol=A/RES/70/1&Lang=E

Wals, A. E. J. (2009). A mid-decade review of the decade of education for sustainable development. *Journal of Education for Sustainable Development, 3*(2), 195–204.

Zwicky, J. (2003). *Wisdom and metaphor.* Kentville, NS: Gaspereau Press.

Bob Jickling has been an active practitioner, teaching courses in environmental philosophy; environmental, experiential, and outdoor education; and philosophy of education. He worked as Professor of Education at Lakehead University after many years of teaching at Yukon College. He continues his work as Professor Emeritus at Lakehead University. His research interests have long included critiques of environmental education and education for sustainable development while advancing formulations for opening the process of remaking education. Jickling was the founding editor of the *Canadian Journal of Environmental Education* in 1996, and together with Lucie Sauvé, he co-chaired the 5th World Environmental Education Congress in Montreal, in 2009. He has also received the North American Association of Environmental Education's Awards for Outstanding Contributions to: Research (2009) and Global Environmental Education (2001). In 2012, he received the Queen Elizabeth II Diamond Jubilee Medal in recognition of contributions to Canada. As a long-time wilderness traveller, much of his inspiration is derived from the landscape of his home in Canada's Yukon.

Assuming the Future: Repurposing Education in a Volatile Age

Stephen Sterling

Abstract The instrumentalist view of the purpose of education manifested in recent decades through the influence of neoliberal ideology, and the set of assumptions that accompany this wave of change and reform, are critiqued from an ecological and humanistic viewpoint. A collective blindness to the global systemic issues that are shaping the near human and planetary future is present both in wider society and in educational systems that can, consequently, be deemed maladaptive to this reality. A deep learning response within educational thinking, policymaking, and practice is required based upon an emerging relational or ecological worldview, already burgeoning in diverse civil society movements. This would allow attention to be brought to generating purposes and assumptions in education aligned to, and able to address, the possibilities of systemic breakdown or breakthrough in global and local systems. Such education is supportive of living in more creative, collaborative, and explorative ways that help assure breakthrough trajectories as the century plays out.

Keywords Neoliberalism · Purpose of education · Global risk · Anthropocene · Participative reality · Breakthrough

S. Sterling (✉)
Centre for Sustainable Futures (CSF), University of Plymouth, Plymouth, UK
e-mail: stephen.sterling@plymouth.ac.uk

© The Author(s) 2017   31
B. Jickling, S. Sterling (eds.), *Post-Sustainability and Environmental Education*, Palgrave Studies in Education and the Environment,
DOI 10.1007/978-3-319-51322-5_3

It is not possible to work in environmental or sustainability education for very long before questions arise about the fundamental purpose of education, particularly now, when educational policy and practice is being trammelled by an economically driven instrumentalism across the globe. After reviewing this trend, the chapter argues for a reimagination of educational purpose that is appropriate and aligned to the unprecedented nature of our times and collective futures.

Asking the Question

"What is the purpose of education?" It was a weekday morning at a large university in the UK, and I stood before some 300 teacher education students. I was there to lecture, but began by putting a big "?" on the a/v screen, and asked this question. There was a hush in the room. No one ventured an answer—well, not for some minutes—and even then the answers were tentative, as if I had asked a trick question. These were students who had just gone through a 3-year degree in educational studies and were about to enter schools as teachers. I was surprised. Their lecturers, dotted around the room, looked a little embarrassed. The question could hardly be more fundamental. Stafford Beer, the systems theorist, put it this way: "the purpose of the system is what it does." And what educational systems do and are for is obvious, isn't it? . . . which renders my question about purpose surprisingly radical—subversive even. Some would see it as superfluous. Because we all know, education is about jobs, and supporting economic competitiveness. Simple. So let's get on with it.

And that is exactly what is happening. Globally, the "education industry" is worth around $5.0 trillion, and growing, and there is something like 200 million students in higher education (HE) worldwide and the number is rising (Verger et al., 2016a). It is big business. And at the same time, the language of business pervades policymaking and discourse in HE. In the UK, a recent Government White Paper, *Success as a Knowledge Economy*, heralds a restructuring of the HE system to make it more open to competition including profit-driven providers, to private funding, and to underline the role of students as consumers. The message regarding the purpose of HE is clear (DBIS, 2016, p. 7):

> Our universities have a paramount place in an economy driven by knowledge and ideas. They generate the know-how and skills that fuel our growth and provide the basis for our nation's intellectual and cultural success.

A counter paper, *In Defence of Public Education* (Holmwood, Hickey, Cohen & Wallis, 2016), mounts a critical response arguing that the White Paper seeks to replace HE with training, and defends the role of universities as havens of critical knowledge essential to a healthy democratic society. However, the instrumental view of educational purpose is echoed in the OECD survey of education trends, where the OECD Secretary-General states (Gurría, 2014, p. 15):

> Education and skills hold the key to future wellbeing and will be critical to restoring long-term growth, tackling unemployment, promoting competitiveness, and nurturing more inclusive and cohesive societies.

The country statistics reflected in this OECD compendium do not reveal the massive influence and reach of the "Global Education Industry" (GEI). Verger et al. (2016b, p. 3) comment on the rise of the GEI, and particularly "the *conception* of education that is increasingly globalised and managed by private corporations" (my italics). The GEI, they maintain, is shaped and enabled by public policymaking which, is itself, "often influenced by the private interests in the GEI as they seek policy agendas, frame policy problems, and refashion regulatory regimes to their advantage" (p. 4).

NEOLIBERALISM AND EDUCATION

This tidal wave—that has swamped older conceptions of education as a public service for the public good—is a relatively recent phenomenon. Over some 30 years, an instrumental view of education has come to dominate, modelled on economic change and the perceived demands of a globalised economy and increasingly globalised culture. This change is not peculiar to the field of education, but "marketisation" and "modernisation" have infiltrated virtually all areas of public life including sports, health, the penal system, policing, and local government over a period of years (Marshall & Peters, 1999).

How education is perceived, conceived, and received is being shaped by a particular view of the world and of people within it. This can be characterised as technocratic, managerialist, economistic, and vocationalist, and is underpinned and energised by an internationally hegemonic neoliberal ideology. Over time, this wave has subtly but powerfully displaced—and is even now drowning out—older (and more educationally defensible), liberal, holistic, and humanistic philosophies regarding the nature and purpose of education. The main vehicle of this instrumental thrust is the "global testing culture" (Smith, 2016) that now envelops not only students, but teachers, lecturers, and entire institutions. Insidiously, "as the global testing culture spreads, cultural models are internalised by actors and the underlying assumptions and values are no longer questioned" (Smith, 2016, p. 11). Similarly, more broadly, Martínez-Rodríguez and Fernández-Herrería (2016, p. 2) argue that a key ideological effect of neoliberal thought is to "deny the possibility of alternative ways of organising both societies in general and education in particular." Smith sees the core assumptions informing the testing culture as positivism and individualism—old ghosts reborn for the late modern age.

Associated educational assumptions that shape this platform include the following:

- Education is a key to economic success.
- The prime purpose of education is to make students employable and the economy more competitive.
- Students are primarily motivated by better employment prospects.
- Students perform best by being tested constantly and being pressed to achieve.
- Quality is assured by competition, metrics, and accountability.
- Cognitive knowledge is prime.
- Students and parents are consumers.
- Values, ethics, emotions, and intuition have little or no place in education.
- The best default pedagogy is "delivery" by experts.
- Educational institutions' performance is enhanced through being in a competitive market.

With such assumptions in place and apparently widely shared, a particular cluster of policies and practices follow logically—to the exclusion of alternatives, and the marginalisation and squeezing of critical and explorative discourses, and of non-conformist practices. Rather, there appears to be a rush to be part of this paradigm, which is increasingly accepted as the "obvious" norm. Here, for example, is a commentary by an American think tank that sets out the "old reality" against the "new reality," under the heading "Old assumptions which no longer fit" (Greer, 2013, p. 5):

The primary goal of college is to produce liberally educated graduates, through a coherent, sequenced curriculum, taught by full-time faculty.

New reality: The major purpose of college is preparation for the workforce, global economic competitiveness, and gaining practical skills through asynchronous, modular content delivery, using part-time faculty and technology. While the laudable goal of providing and acquiring a "liberal" education is alive, it is not as healthy as it once was, and from the view of many colleges and students, may be on the way to becoming an anachronism.

What we seem to have here is an unquestioning narrowing, a diminution—a debasement even—of the meaning and purpose of education, but also, and more gravely, of what it means to be human. A critical blog puts it eloquently (Strauss, 2013):

> Over and over again, reformsters suggest that the only real purpose of an education is to prepare one for work. You get an education so that you can become useful to your future possible employers. That's it. That's all. Everything that is beautiful and loving and glorious about human life, everything that resonates in our connections to each other and the world around us—none of that matters in education. The measure of whether a subject should be taught is simply, "Will this help the student get a job?" Learning about everything that is rich and joyful and rewarding in the human experience, everything about learning to grow and understand and embrace who you are as a human being and how you make your way in the world—that's all stuff you can do in your free time, I guess, if you really want to.

A key part of the reformist landscape is standardisation, a narrowing of what counts as curriculum, homogenisation and testing, in the name of accountability and comparison. This is viewed as the only legitimate measure of quality—the human consequences of this regime notwithstanding. In October 2015, a report showed half of all teachers in English schools were considering leaving the profession citing the combined negative impact of the accountability agenda on teacher workload and morale (Boffey, 2015). Further, a poll in 2016 (Press Association, 2016) found that 89 per cent of teachers polled claimed that exams and testing were resulting in increasing level of student stress and self-harm.

Such negative effects of the limited purpose of education in the era of the late-modernist agenda are not an intended part of the script, but they are perhaps inevitable in a system which is so narrowly prescribed and relentlessly driven. Because, for all the common claims for "new policies responding to the needs of a rapidly changing world," the model is still one of command

and control, which as Chapman (2002, p. 12) remarks "inevitably fails within complex systems, and alienates people by treating them instrumentally."

We can invoke here the notion of "systems failure" (Peters, 1999; Chapman, 2002). According to Peters (1999, p. 124), failure can be considered to be of four types: objectives not met, undesirable side effects, designed failures, and inappropriate objectives. Criticism of education—particularly in political debate—often centres on the first meaning, but the additional criteria have wider resonance. I have already touched on undesirable side effects of instrumentalism, but there is a more pressing issue that relates to the last two criteria. This concerns a prevailing and baffling blindness to the global existential crisis, to the global threats and "grand challenges" that are already affecting, and will increasingly dominate, all people's lives—particularly those of the younger generation. Hence the title of this chapter: *Assuming the Future*.

The future seems to be regarded by the mainstream as some sort of constant, assured, and stable, whilst the normal business of educating/training for jobs and any kind of economic growth proceeds untrammelled and unbothered by notions of: resource depletion and competition, poverty and growing inequity, marginalisation of minorities, spreading fundamentalism, extremism and terrorism, the implications of the march of bioscience and robotics, species loss and plummeting biodiversity, climate change, food security, wars and civil unrest, the risk of global pandemics, and so on. The stark reality is that we live in an age of unprecedented systemic global risk, and this is corroborated by leading academics and think tanks (see, e.g., Armstrong & Pamlin, 2015; or WEF, 2016). Yet, in education (as other sections of society) there is a strange ignorance—ignore-ance—of *context*, which is hard to square with the slew of international reports in recent years about global challenges, well summarised in Al Gore's (2013, p. xv) assertion:

> There is a clear consensus that the future now emerging will be extremely different from anything we have ever known in the past.... There is no prior period of change that remotely resembles what humanity is about to experience.

THE ANTHROPOCENE, MALADAPTATION, AND RESPONSE

So how about a sober reality check? Every one of us is now living in the age of (what is increasingly referred to as) the Anthropocene—unprecedented times when humanity, as the prime agent of planetary change, bears a deep moral responsibility to the Earth, to future generations of humans, and to all other species—not least as the stability of planetary systems is now under real threat. As Johan Rockström (2015) of the Stockholm Resilience Centre states, a "profound new risk can be added to the conventional concerns of dwindling resources and local pollution: human action could push the Earth system to abrupt and irreversible shifts of the planetary ecosphere." In response, there is certainly evidence of an arising, slow-burning, yet accelerating awakening to the need for a fundamental shift of the human trajectory. For example, Renner (2015, p. 170) in a recent Worldwatch Report, argues that:

> the challenge for humanity today is no longer anything like what it faced in the 1960s and 1970s, when developing pollution abatement technologies and lessening the degree to which resources were wasted provided a more-or-less adequate answer to the most pressing problems of the day. The world now needs to adopt solutions that change the entire system of production and consumption in a fundamental manner, that move societies from conditions of energy and materials surplus to scarcity, and that develop the foresight needed to recognize still-hidden threats to sustainability.

More radically, the powerful notion of the Great Transition (first proposed by the ecological economist Kenneth Boulding in the 1960s) towards a world of universal human solidarity, well-being, and respect for nature is gaining traction and attention. This elaborates alternative scenarios indicating both desirable and dystopian pathways into the future and thus highlights the possibility—within a relatively narrow window of time—of conscious choice (Raskin et al., 2002). There is then, a need for *unlearning*, *re-learning*, and *new learning* as a necessary response to a deeply changing reality. Bizarrely, however, the tasks of radical reassessment and critical reflection this requires, hardly figure in mainstream educational discourse let alone most political and economic debate. Yet voices for change are becoming more insistent. For example, Escrigas (2015) appeals to universities, arguing they should "learn to read reality," and "understand the wider impacts of their actions and the costs of what they are *not* doing at a time when societal transition is urgently needed."
As I have written earlier (Sterling, 2009, p. 19):

> The paradox of education is that it is seen as a preparation for the future, but it grows out of the past. In stable conditions, this socialization and

replication function of education is sufficient: in volatile conditions where there is an increasingly shared sense (as well as numerous reports indicating) that the future will not be anything like a linear extension of the past, it sets boundaries and barriers to innovation, creativity, and experimentation.

Even something as mainstream as the employment agenda in HE indicates a shortfall in contextual awareness. This is somewhat ironic, as the whole area of employability has become a kind of *raison d'etre* and expressed major purpose for HE—a justification for marketised, "business-facing" education. Yet virtually missing in mainstream education and its employ-ability debate is recognition of "technological unemployment"—the relentless trend towards automation of jobs across the spectrum, and the massive implications this may have for society (Avent, 2016). The UK Government White Paper on HE (introduced above)—which is entirely oriented towards employability and growth—makes no mention of this vital topic, despite insisting in its Executive Summary (2016, p. 7) that "we must be ready for the challenges of the future."

So we might say that with regard to the specific topic of employment, or to wider global trends, the education system has grown maladaptive. In other words, this learning system is not *itself* learning. It is insufficiently responsive to the profound socio-economic and ecological shifts that characterise the global systems within which it is embedded. Despite evidence of an increasing number of initiatives, programmes, and research projects which really *are* making a positive difference to the world, on balance, the assumptions governing most education policies and practices do not acknowledge, reflect, or respond to the highly systemic, risk-laden and challenge-beset present and future that graduates are entering. So judged in terms of net benefit, the effects of such education may often be more harmful than remedial. Tom Bentley argues that grand organising narratives—whether, for example, Marxism, Christianity, or indeed the market—have provided an imposed, definitive, closed account of what matters and what it means for our lives. He maintains these need to be replaced with a far more open and flexible view, where solutions are not preordained but generated, based on "our capacity to behave intelligently and to learn" (Bentley, 1998, p. 172). This is a plea for adaptive or second-order learning to be engendered involving all levels—individuals, institutions, communities, and indeed educational systems. Such learning questions redundant, dysfunctional assumptions and values, and rather, develops values and ways of seeing that are resonant with, appropriate to, and able to address the conditions of complexity, difficulty, uncertainty, hopes, and fears that are increasingly the real-world experience for vast numbers of people.

Evidence of a response that questions the bounds and norms of the neoliberal paradigm in education is growing. For example, UNESCO commissioned a report *Rethinking Education* which argues for a holistic and humanistic renewal of education globally as "a common good" (UNESCO, 2015). This report counters the prevailing commodification of education as a private good. In fact, UNESCO has been suggesting the need for a "new vision" of education for some time. What is often missing in such calls is first, a sufficient critique of the dominant cultural worldview and associated educational policies and practices, and second, the articulation and elaboration of a necessary alternative which might convincingly underpin such a vision. We lack a widely shared alternative paradigm of educational philosophy, policy, and practice which is at once humanistic and ecological, aligned and responsive to the complex social-ecological trends and risks now manifest within our Anthropocene times.

Of course, many "education for change" movements have attempted to put alternatives into practice over the past 30 years or so. Among the lists of "adjectival educations," notably environmental and sustainability education—however conceived—have made worthy progress in carving out spaces for thought, exchange, and practice of alternatives. Without entering the choppy and extensive waters of the Education for Sustainable Development (ESD) debate, I would maintain that the banner of "ESD" has helped precipitate positive work internationally through providing a kind of legitimised door into the mainstream. However, I also acknowledge that work presented under this label is often subject to accommodation and neutering by the same mainstream, particularly where any more radical notions dare raise their heads. Labels cannot be avoided, but we should always be aware that they can delimit, divide, and confuse as much as clarify and communicate.

PARTICIPATIVE REALITY, EDUCATION, AND SOCIETAL BREAKTHROUGH

What is of critical importance is the nature of the assumptions and values that inform educational thinking, policies, and practices—although the tricky issue of terminology is ever present. We have an urgent and vital task: to develop a sane and robust alternative to the neo-liberal view of education. But this must clearly arise from a very different starting point as regards an informing worldview. In my own work, the environment has always been a starting point and grounding, inviting identification and sensibility beyond the self to a far greater reality, and this has underpinned my vocation in environmental and sustainability education. Yet the word "environment" is problematic. I was struck years ago, by an idea in a book by Kenneth Boulding (1978, p. 31):

> We must look at the world as a whole... as a total system of interacting parts. There is no such thing as an "environment" if by this we mean a surrounding system that is independent of what goes on inside it.

The very idea of the environment as some sort of *separate* reality is delusional, and stems from a sense of disassociation deep within our psyche relating to the potent and ingrained Western intellectual legacies of mechanism, dualism, objectivism, reductionism, and so on. In my view, it is this myopic cultural and perceptual orientation that lies at the heart of the global existential crisis. As psychiatrist Iain McGilchrist states, "the kind of attention we pay actually alters the world: we are, literally, partners in creation. This means we have a *grave* responsibility..." (author's italics) (2009, p. 5). Far from being detached and unaffected observers, we are—unavoidably—participants within a greater whole, a participative reality that necessitates an essentially *relational* worldview and episteme. For me, this means that environmental and sustainability education have never been, and cannot be, ends in themselves, contained and complete. Rather, they imply and act as outriders or a vanguard for a necessary deeper shift in educational culture. They point to a need for an ecological educational paradigm appropriate to the world we inhabit, and the critical conditions we have created (see Sterling, 2001).

How do we pay fuller attention to the more-than-human world, to its wonder, its beauty, its suffering, and to the dignity, needs, and wholeness of every person? How do we maintain openness, value emergence, creativity, and explorative learning as we create a saner, more liveable future in conditions of volatility and contingency? How do we become more mindful of our own thinking, assumptions, and behaviour that can move us towards a deeper self-knowing? Mainstream educational thinking and practice, where it has become narrower and shallower under the influence of the neoliberal agenda, is deleterious to this necessary journey of awareness and transition. Any educational paradigm worthy of the name needs to support a remedial movement in three interrelated areas of human knowing and experience to transcend dysfunctional worldviews and help heal our world. This can be summed up as: a broadening of *perception* (the affective dimension), a shift towards relational thinking or *conception* (the cognitive dimension), and manifestation of integrative *practice* (the intentional dimension) towards well-being. In sum, an extended, inclusive and participatory epistemology, a relational ontology, and an integrative praxis can nurture a deep ecological sense of what it is to be human at this most challenging of times, through changed educational thinking, policy, and practice.

Reliable futures scanning and scientific reports indicate the possibility—and some, the probability—of global breakdowns or collapse scenarios in this current century through a variety of stresses including economic meltdown, technological vulnerability, social upheaval and mass refugee movements, disease pandemics, food and energy shortfalls, disruption of ecosystems, climate change, and so on. But these are not some far-off scenarios. Already, global media coverage of localised manifestations of such phenomena carries an uneasy sense of immediacy to otherwise comfortable audiences.

Homer-Dixon (2006) makes an important distinction between societal breakdown and collapse. Whilst both produce a radical reduction of complexity in a system, and thereby reduce future options, collapse is potentially catastrophic. Breakdown, however, allows potential for recreation of social and other human systems. He argues that the fundamental challenge the world faces is to anticipate and allow for breakdown in a way that does not lead to collapse, but to renewal. And thereby to break*through*. We have a choice: "our challenge isn't to preserve the status quo but rather to adapt to, thrive in, and shape for the better, a world of constant change" (Homer-Dixon, 2006, p. 266). Again, there is a clear message for those that purport to educate for the future (and surely all education is about the future?). I made this necessary point some 20 years ago, and it is worth repeating here (Sterling, 1996, p. 26):

> Whether the future holds breakdown or breakthrough scenarios . . . people will require flexibility, resilience, creativity, participative skills, competence, material restraint and a sense of responsibility and transpersonal ethics to handle transition and provide mutual support. Indeed, an education oriented towards nurturing these qualities would help determine a positive and hopeful breakthrough future.

And so we come back to the purpose of education. Changing the purpose or goals of a system has the power to effect systemic change, secondary only to paradigm shift (Meadows, 2009). After decades of arguing for a change of the dominant educational paradigm towards something more holistic, systemic, humanistic, and ecological (Sterling, 2003), I fully understand that the realisation and internalisation of different educational paradigms by individuals, institutions, and educational communities is extraordinarily challenging. But a change of *purpose*—or embrace of additional purpose in the first instance—is possible at micro, meso, and macro levels. This can be a harbinger of a deeper cultural shift, especially when aligned with, and connected to, growing progressive and reconstructive movements in civil society (Martínez-Rodríguez & Fernández-Herrería, 2016) inspired by values reflecting ecological integrity, social justice, and ethics—such as are represented by the Earth Charter.

CONCLUSION

There is overwhelming scientific evidence that the planetary and human future involves grave risks in this century. As Rockström (2015) states, "Our historical condition does, whether we like it or not, change everything." Indeed. Everything—except, it would seem, the very system that is supposed to prepare people for life and their individual and collective futures. Whilst, encouragingly, there are a growing number of laudable exceptions in policy and practice around the world, for the most part educational systems—almost perversely—pay insufficient heed to the deep challenges that face us all collectively. The role and purpose of education can no longer be preparation for an assumed stable future and "business as usual," but a nurturing of individual and collective potential to live well and skilfully in an already complex and volatile world, towards human and planetary betterment.

Many years ago, I heard the distinguished British environmental educator, and "gentle anarchist," Colin Ward speak. One line stuck with me: "What matters is the quality of your assumptions," he said. Let us assert educational thinking and practice built on a keen and renewed sense of purpose and assumptions that arise from our common humanity and commitment to a safer, kinder, and flourishing world and planet. A reimagined education appropriate for this volatile age is not constrained, prescribed, and judgmental—but supportive, inclusive, developmental, trust-building, explorative, critical, creative, experimental, holistic, transformative, and diverse according to local and individual need and potential. Let it flourish wherever it may: in this vital endeavour, environmental and sustainability education continue to have crucial roles as pathfinders and pathmakers.

REFERENCES

Armstrong, S., & Pamlin, D. (2015). *Global challenges—12 risks that threaten human civilisation*. Global Challenges Foundation, Future of Humanity Institute, Oxford: Oxford University. Retrieved from http://globalchallenges.org/wp-content/uploads/12-Risks-with-infinite-impact.pdf

Avent, R. (2016, October 9). Welcome to a world without work. *The Observer*, pp. 37–39.

Bentley, T. (1998). *Learning beyond the classroom—Education for a changing world*. London: Demos/Routledge.

Boffey, D. (2015). Half of all teachers in England threaten to quit as morale crashes. Retrieved from http://www.theguardian.com/education/2015/oct/04/half-of-teachers-consider-leaving-profession-shock-poll

Boulding, K. (1978). *Ecodynamics—A new theory of social evolution*. London: Sage Publications.

Chapman, J. (2002). *System failure*. London: Demos.

Department for Business, Innovation and Skills (DBIS). (2016). *Success as a knowledge economy: Teaching excellence, social mobility and student choice*. Retrieved from https://www.gov.uk/government/uploads/system/uploads/attachment_data/file/523396/bis-16-265-success-as-a-knowledge-economy.pdf

Escrigas, C. (2015). *A higher calling for higher education*. Boston, MA: Tellus Institute. Retrieved from http://www.greattransition.org/publication/a-higher-calling-for-higher-education

Gore, A. (2013). *The future*. New York: W.H. Allen.

Greer, D. (2013). *New assumptions and new solutions for higher education reform*. Center for Higher Education, Strategic Information and Governance, NJ: The Richard Stockton College of New Jersey. Retrieved from http://intraweb.stockton.edu/eyos/hughescenter/content/docs/HESIG/Working%20Paper%201%20for%20Website.pdf

Gurría, A. (2014). Editorial—Education and skills for inclusive growth. *Education at a glance 2014 OECD indicators*. Paris: OECD Publishing. Retrieved from https://www.oecd.org/edu/Education-at-a-Glance-2014.pdf

Holmwood, J., Hickey, T., Cohen, R., & Wallis, S. *The alternative white paper for higher education—In defence of public education: Knowledge for a successful society*. Retrieved from https://heconvention2.files.wordpress.com/2016/06/awp1.pdf

Homer-Dixon, T. (2006). *The upside of down—Catastrophe, creativity and the renewal of civilisation*. London: Souvenir Press.

Marshall, P., & Peters, M. (1999). *Education policy*. Cheltenham: Edward Elgar Publishing.

Martínez-Rodríguez, F., & Fernández-Herrería, A. (2016). Is there life beyond neoliberalism? Critical socio-educational alternatives for civic construction. *Globalisation, Societies and Education, 14*, 1–12. http://www.tandfonline.com/doi/ref/10.1080/14767724.2016.1195726 doi: 10.1080/14767724.2016.1195726

McGilchrist, I. (2009). *The Master and his emissary: The divided brain and the making of the western world*. New Haven: Yale University Press.

Meadows, D. (2009). *Thinking in systems—A primer*. London: Earthscan.

Peters, G. (1999). A systems failures view of the UK National Commission into Higher Education Report. In R. Ison (Ed.), *Systems Research and Behavioral Science, 16*(2), 123–131.

Press Association. (2016). Primary school pupils 'driven to self-harm amid tests and social media stress'. Retrieved from http://www.aol.co.uk/news/2016/04/04/primary-school-pupils-driven-to-selfharm-amid-tests-and-social-media-stress/

Raskin, P., Banuri, T., Gallopin, G., Gutman, P., Hammond, A., Kates, R., & Swart, R. (2002). *Great transition: The promise and lure of the times ahead*. Boston: Stockholm Environment Institute/Tellus Institute.

Renner, M. (2015). The seeds of modern threats. In L. Mastny (Ed.), *State of the world 2015: Confronting hidden threats to sustainability* (pp. 3–17). Washington, DC: Earthwatch Institute, Island Press.

Rockström, J. W. (2015). *Bounding the planetary future: Why we need a great transition*. Tellus Institute. Retrieved from http://www.tellus.org/pub/Rockstrom-Bounding_the_Planetary_Future.pdf

Smith, W. (2016). *The global testing culture—Shaping education policy, perceptions, and practice*. Oxford: Symposium Books.

Sterling, S. (1996). Education in change. In J. Huckle & S. Sterling, (Eds.), *Education for sustainability*. London: Earthscan.

Sterling, S. (2001). *Sustainable education—Re-visioning learning and change, Schumacher briefing*. Dartington: Green Books.

Sterling, S. (2003). *Whole systems thinking as a basis for paradigm change in education: explorations in the context of sustainability*. PhD diss., Centre for Research in Education and the Environment, University of Bath. Available at: www.bath.ac.uk/cree/sterling/sterlingthesis.pdf

Sterling, S. (2009). Towards sustainable education. *Environmental Scientist, 18*(1), 19–21.

Strauss, V. (2013). Five bad education assumptions the media keeps recycling. Retrieved from https://www.washingtonpost.com/news/answer-sheet/wp/2013/08/29/five-bad-education-assumptions-the-media-keeps-recycling/

UNESCO. (2015). *Rethinking education—Towards a global common good?* Paris: UNESCO.

Verger, A., Lubienski, C., & Steiner-Khamsi, G. (2016a). *The rise of the global education industry: Some concepts, facts and figures.* Retrieved from www.educa tionincrisis.net/blog/item/1308-the-rise-of-the-global-education-industry-some-concepts-facts-and-figures

Verger, A., Lubienski, C., & Steiner-Khamsi, G. (2016b). The emergence and structuring of the global education industry—Towards an analytic framework. In A. Verger, C. Lubienski, & G. Steiner-Khamsi (Eds.), *The world yearbook of education 2016: The global education industry.* London: Routledge. Retrieved from https://www.book2look.de/embed/HNoMrtccfl&euid=56938807&ruid=56938806&referurl=www.routledge.com&clickedby=H5W&biblettype=html5

World Economic Forum. (2016). *The global risks report 2016* (11th ed.). Geneva: WEF. Retrieved from https://heconvention2.files.wordpress.com/2016/06/awp1.pdf

Stephen Sterling is Professor of Sustainability Education, Centre for Sustainable Futures (CSF) at the University of Plymouth, UK. His research interests lie in the interrelationships between ecological thinking, systemic change, and learning at individual and institutional scales to help meet the challenge of accelerating the educational response to the sustainability agenda. He has argued for some years that the global crises of sustainability require a matching response from the educational community, together with a shift of culture towards educational policy and practice that is holistic, humanistic, and ecological. A former Senior Advisor to the UK Higher Education Academy on Education for Sustainable Development (ESD) and current National Teaching Fellow (NTF), he has worked in environmental and sustainability education in the academic and NGO fields nationally and internationally for over three decades. He was a member of the UNESCO International Reference Group for the UN Decade on Education for Sustainable Development (2005–2014), and he is currently co-chair of the International Jury for the UNESCO-Japan Prize on ESD. Widely published, his first book (co-edited with John Huckle) was *Education for Sustainability* (Earthscan, 1996), and this was followed by the influential Schumacher Briefing *Sustainable Education—Re-visioning Learning and Change* (Green Books 2001). He also co-founded the first masters course in the UK on sustainability education (at London South Bank University), and led the WWF-UK project on systems thinking *Linking Thinking—new perspectives on thinking and learning for sustainability.*

His work at Plymouth involves developing strategies to support whole institutional change towards sustainability. Other books include (with David Selby and Paula Jones, Earthscan 2010) *Sustainability Education: Perspectives and Practice across Higher Education* and (with Larch Maxey and Heather Luna) *The Sustainable University—Process and Prospects,* published by Routledge in 2013.

PART II

Critique and Proposition

Saying Yes to Life: The Search for the Rebel Teacher

Sean Blenkinsop and Marcus Morse

Abstract The chapter starts with suicide and ends in rebellious possibility. We begin by highlighting Albert Camus's consideration of suicide, and in particular his assertion that in the act of choosing not to exercise our ever-present radical freedom to commit suicide there exists both a negation, saying no to suicide, and an exaltation, of saying yes to life. Camus's purpose in this is to have us actively consider why we are choosing to stay alive and, as such, rebel against the possibility of suicide—and with purpose, choosing to say yes to life. We then focus on the distinction Camus draws between revolution and rebellion to allow us to more deeply explore his concept of the rebel and the shared role that negation and exaltation play therewith. By exploring Camus's existentialist concept of freedom in order to name both a particular negation and exaltation for our times, as he was doing for his own time, we meet Camus's challenge to consider, name, and act upon that which we choose to say yes to. The chapter concludes with an exploration of implications for

S. Blenkinsop (✉)
Simon Fraser University, Burnaby, Canada
e-mail: sblenkin@sfu.ca

M. Morse
Outdoor and Environmental Education, La Trobe University, Melbourne, Australia
e-mail: M.Morse@latrobe.edu.au

© The Author(s) 2017   49
B. Jickling, S. Sterling (eds.), *Post-Sustainability and Environmental Education*, Palgrave Studies in Education and the Environment,
DOI 10.1007/978-3-319-51322-5_4

environmental educators who want to adopt, and build upon the initial work presented in this chapter in pursuit of becoming creative, rebel teachers.

Keywords Education · Environmental · Ecological · Sustainability · Camus · Existential · Pedagogy

Albert Camus, the French-Algerian philosopher, writer, director, and Nobel Prize winner died at 46 in a car crash. Yet before this tragic accident he produced a body of work that continues to engage, provoke, and challenge Western thought. His is a powerful voice for justice coupled with the life lived as an activist. It is likely that his history as a child in a single parent home and growing up poor in a French African colony influenced him in these directions. Throughout his career, Camus tended to write novels and philosophical essays in tandem and it is from one of these pairings that this chapter arises: *The Plague*, the story of how a town and its people respond to an outbreak of the bubonic plague, and *The Rebel*, a series of philosophical essays exploring how individuals and communities might change their culture. Both books (first published in 1947 and 1951, respectively) are deep explorations into what humans should/could/might do in the face of the seemingly insurmountable challenges[1] both to respond to the immediate problems and to work to create a different culture such that those same problems did not reoccur.

Underneath these explorations is the question that Camus places at the centre of his work and of all philosophy. A question that he suggests is really the only one of importance for all philosophers—suicide. Now at first blush this may seem depressing, potentially nihilistic, and too narrow in scope for philosophy and our purposes of opening new terrain for environmental educators beyond the well-trodden, at times lifeless ground of sustainability and stewardship. But Camus, in his own blunt and provocative way, is asking a significantly more expansive question, one that parallels the classic Socratic/Platonic question of the good life, only through the methods of a rebellious existentialist and more secular mind of a twentieth century theorist. As humans we have the ability to freely commit suicide, to exercise "our radical freedom," and thus the question is why, on this day or the next, do each of us individually decide not kill ourselves. Why are we choosing to live, and by extension, what is that choice saying about what we think is of value and about how a life should/

could be lived. Camus's point is to have us actively consider why we are staying alive and, as such, rebel against this possibility of suicide and with purpose, say, "yes" to life. It is this combination of negation, saying no to suicide, and exaltation, saying yes to life, that informs Camus's concept of the rebel and that might influence environmental educators who see the need for a substantial change in pedagogy, curricula, and even educational culture to better think through how to enact this freedom to which, for Sartre, we are all condemned.

In *The Rebel*, Camus (1956) sets out to understand the times in which he lived and examine the crime of his time, the extermination of tens of millions of people. He examines the apparent acceptable logic of such a crime and takes up the challenge to understand how this crime could occur and be accepted by so many; "on the day when crime dons the apparel of innocence—through a curious transposition peculiar to our times—it is innocence that is called upon to justify itself" (p. 4). Camus begins with the question of suicide because without a workable formulation for suicide, then murder cannot be critiqued.[2] The consideration of suicide is used as the starting point because its avoidance, or for Camus, its active negation (one is choosing not to kill oneself for express reasons) involves a necessary affirmation and decision to say yes to life. And for Camus, it follows, that if we are to recognize and assert that there is something good in life, something worth saying yes to then the same is true for others; "from the moment that life is recognised as good, it becomes good for all men. Murder cannot be made coherent when suicide is not considered coherent" (p. 6). And we shall see the implications of this mutual involvement in saying yes to life and freedom as the chapter progresses.

So it is this question of suicide that ignites the thinking in this chapter and not in the same way Camus initially intended, but in a way that reflects our current times and of which he would likely have approved. If, and the scientific and non-scientific evidence seems quite convincing, we are in the midst of an environmental crisis of global proportions then, to echo Camus, we suggest that the only real question for environmental educators is this question of suicide. Not so much our own ability to commit it, although that remains critically important for each individual, but what appears to be the direction for our species: we are committing mass-suicide.[3] The question for this chapter then is what can we learn from Camus that might help us respond to this crisis of suicide and find ways for this species[4] to say "yes to life?"

In this chapter we focus on the distinction Camus draws between revolution and rebellion to allow us to more deeply explore his concept of the rebel and the shared role that negation and exaltation play therewith. We explore Camus's existentialist concept of freedom in order to name both a particular negation and exaltation for our times, to meet Camus's challenge of considering, naming, and acting upon that which we choose to say yes to. The chapter includes a brief exploration of implications for environmental educators who want to adopt, and build upon, the initial work presented in this chapter in pursuit of becoming creative, rebel teachers.

PART 1: REVOLUTION VERSUS REBELLION

Camus, in *The Rebel*, traces the story of rebellion and revolution through history, the arts, and a study of metaphysics. His goal is to paint a picture of who the rebel is, how rebellion works, and to articulate the pitfalls of revolution. For Camus, those who revolt and the revolutions that result are not and have never been overly successful. This is for several reasons, two of which are important here. First, revolutions tend to be about the destruction and annihilation of an entire current system, culture, or way of being and therefore lack, in the way of nihilism, a positive impetus, replacement, or possibility with respect to the future. And the second reason is because revolutionaries forget that any change must start from where things are. For Camus, not only does revolution usually begin with an absolute negation, say of an entire class or culture, but also attempts to respond by assuming that living, breathing humans can leap, at will, directly from one paradigm/way-of-being to another with no bridge, support system, or intermediary structures. For Camus this presumed "absolute flexibility" is flawed and arises because the revolutionary theorizers have forgotten that these are real people, always and already in the world and, we add, they are ignoring the educational nature of cultural change.

Camus (1956) contrasts revolution with the idea of rebellion that involves an understanding that change is much more messy and conflicted—but also more grounded and practical. He suggests that rebellion "starts with a negative supported by an affirmative" (p. 251). Change happens not as complete overthrow of history or metaphysics but as an ongoing paradoxical project of negation and exaltation at the same time, "it says yes and no simultaneously. It is the rejection of one part of existence in the name of another part, which it exalts" (p. 251). In a similar way to the rejection of suicide, whereby if we decide to say yes to life it is because there is something good worth living for. For Camus, if we decide to rebel—it must be because we have found something worthwhile in our culture, environment, and/or each other worth fighting for. In both cases, though, these exaltations are not given; rather, they are arrived at via the act of living within the world. The challenge is to select the best negation and exaltation for the given moment in an overall rebellion. For, as Camus points out, rebellion is a process with a distant goal of a better way, what he calls unity, but it must honour those engaged in the process of change that cannot leap from here to unity in a single bound: "we can act only in terms of our own time, among the people who surround us" (p. 4). In recognizing the actual state of the human, Camus is acknowledging the reality of being human/alive and returning what he calls our dignity and beauty to us; "but the affirmation of a limit, a dignity, and a beauty common to all men [sic] only entails the necessity of extending this value to embrace everything and everyone and of advancing toward unity without denying the origins of rebellion" (p. 251). It is this recognizing of the other as unique, as situated, and as limited which allows for a genuine noticing of the other. And this noticing brings with it the proffered recognition of this other's subjectivity thereby conferring the dignity of the individual so necessary to Camus's formulation of justice and freedom.

For Camus, most of the revolutions one thinks of throughout politics, history, or art result in everything coming to a standstill as the previous is completely negated and then, eventually, there is a return to the *status quo* for there was nothing positive upon which to build, and the magnitude of the change required is too great. In response to this negating destruction Camus wants to posit the rebel and their rebellion. The rebel acts creatively through rebellion not to the entire system or culture but consciously names and negates the parts of it that they deem to be most problematic, troublesome, and unjust at this time, while also actively exalting something else which might replace that which has been negated. We must, then, engage in the process of understanding a suitable negation and exaltation for our times.

Take the sentiment being echoed by Hay (in Newton & Hay, 2007):

We are called the anti-folk. Anti—this, that, everything. But we are not. We are for, not against. For a tangible, physical place. For the riotous, loving dance of life. For the beauty that will settle anew upon the island when the present horrors pass. (p. 22)

Here the author is responding to the way in which some sectors of the culture in Tasmania are talking about the environmental movement. His point is that there is an assumption and an attempt by the government, press, and business interests of the time to position environmentalists as being anti-everything, as being dreamers, nihilists, or utopians. Or as Camus would say, revolutionaries, whereas Hay is in fact claiming place, life, and beauty as the things which he and other environmental thinkers are, to use Camus's term, *exalting*. And for Camus (1956) it is this combination of negation—the active naming of those things which one wants rid of in the current system—and exaltation—the active naming of those things that are of value, significance, and importance—that makes up the way of rebellion. Rebellion is about changing that which is into that which is desired without making the gap impossible to jump across for real humans. For Camus this is an in-formal rule, or a guide, which he thinks "can be best described by examining it in its pure state—in artistic creation" (p. 252). Thus rebellion is a creative undertaking that makes everyone a part of a process that honours and wants to raise each individual, that understands this as a shared endeavour, and that sees this change as a process already in play even with the first act of saying "No." An act that says no to a specific and identifiable component of the culture, community, or way of being rather than a grandiose negation of its entirety. In this case, the first "No" might be to our shared suicide and then to the critical components of our culture that are so clearly contributing to our deaths. But not without at the same time clearly naming, as Hay does above, that which we are exalting, that which we might seek to build upon as we move away from that which we negate. This step of choosing the first sets of negations and exaltations in this process of change is rife with challenges. This chapter humbly proposes to explore these challenges a little further and offer a couple of possibilities.

PART 2: SEEKING NEGATION AND EXALTATION

At this point, following Camus's suggestions, it would behove us to locate, or settle upon, the initial negations and exaltations that might propel the process. And this process must involve all affected, everyone and everything, to rise with the tide. With this in mind, coupled with the fact that we are at a different historical time facing a different crime, we are proposing a further radical postulate: that everywhere, in every argument in this discussion we push to think of humans and all other living beings.

Thus, in Camus's discussion about rebellion we are not solely talking about humans. And, in the upcoming exploration of freedom, it is implied that freedom is sought not just for all humans (including the marginalized, colonized, and victims of myriad violences), but also for all living more-than-humans.[5]

We are seeking the potential starting points for a rebellion against the non-environmental, colonizing, imperialistic cultural ethos of the North and West, and its trajectory towards suicide for humans and murder of myriad other species. In particular, we are seeking rebellious starting points within the educational system through which the troublesome culture is sustained and supported. The hints that Camus provides are that, (a) we are not seeking absolute negation, that is, that Western culture is all bad, (b) we need to bear in mind this work is going to be done by real beings who are immersed in real places, (c) this work needs to be done with a view to all, and (d) given the limits of humans we must make the specific negation recognizable and workable while also finding the good, even great, components that can be exalted, drawn upon, even, with a nod to Val Plumwood, actively foregrounded. We have a clear first step in the process, negation—a no to suicide, and some indicators from Hay of possible exaltations—place, life, and beauty—but these seem both fairly generic and hard to enact. As environmental educators, naming the crisis for what it is, a suicide, and using that as impetus might be an important first pebble into the pond. Yet it appears we need something more grasp-able, more workable if we are really going to take this project forward, and it is here that we turn back to Camus's idea of freedom because it reverberates with a sense of interconnectedness that reminds us of recent eco-philosophical work.

PART 3: FROM FREEDOM TO FLOURISHING

Although our impetus for exaltation is taken from freedom in our time, Camus again offers provocations to assist. Quite early in Camus's 1949 play *The Just*, we hear a character make a claim that freedom is not an individual achievement, prerogative, even reality but that it is connected to and con-tingent upon others, or as Sartre (1992) suggests; "I am not free unless others are as well" (p. x). This interconnection is important to our discus-sion here in at least two ways. The first is that freedom only exists if there are others who recognize one as something other than an object.[6] To be a chooser in an automatic, pre-programmed, solely instinctual world empty of others, to recognize and share this ability, is to create art in a world with no audience or to make a noise where nobody hears. But, this presence of others is not just about having an audience, but is about being recognized by, and implicated in, the processes of enacted freedom. For if I recognize an other as being free, and this must happen in order for my own freedom to exist in turn, then I have the responsibility to weigh my decisions, actions, and ways of being in light of their possibilities, just as their decisions, freely made, impact my own possibilities in potentially limiting/restricting or opening/expanding ways.

The second way in which the other is necessary to freedom is in its coupling with responsibility such that freedom is about the ability to choose to act and not act, to be and not be, such that every actor/being has the same opportunity and range of possibility. For by extension, if I am not free in a world of automated objects so, too, am I not free if I am the only one able to engage in a full range of possibility in a world of deeply limited others. For what does it mean if I alone am the one who can create what I am? Or if in that process of self-becoming I have made it impossible for others to do the same? Or if only a small portion of living beings, the portion that are not dis-enfranchised, colonized, or marginalized in body or spirit, can exercise it? For Camus this second point is positioned within a social justice conversation and yet Camus himself has opened the door to including not just humans in this project of freedom but also more-than-humans. What if in exalting the possibility of freedom for all we actively include the more-than-humans?

The exaltation we propose, then, is mutually dignified flourishing. This is a rewording of freedom for all, but it makes clear that the freedom being sought is one that comes coupled with responsibility to oneself, to others, and to one's larger community (read ecosystem). The addition of "dignified" reminds us that this project of rebellion is about living and letting live in order to create who we are, and that in losing this ability any specific living thing loses the dignity implicit in being, their ability to self-create, and to be recognized as such beyond simply existing for others or as an unchanging object.

The negation we propose, while less developed in this chapter, is there by implication: individualistic anthropocentrism. This builds upon the message of interconnection and the idea that one is never free, self-becoming, or in rebellion alone. The addition of anthropocentrism may be a surprise, but for the reader willing to recognize that the more-than-human world has created a place where humans can actually exist, that it is made up of myriad unique individuals doing their own thing, and that it is more than simply an amorphous backdrop of objects, then this might in fact be an obvious first negation. We are proposing that we actively begin to undo everything that places the human at the centre and alone while extending the idea of connection, dignity, and increasing possibility for all.

CONCLUSION: ENVIRONMENTAL EDUCATION AND THE REBEL TEACHER

Camus's novel *The Plague* is the story of a town facing death. The plague has arrived, people are dying, and there appears to be little that can be done in response. And yet, in the book, we come across myriad people each responding in their own way to this seemingly overwhelming crisis. Most commentators suggest that Camus is commenting upon French resistance to the Nazi regime as it expanded across Europe, but even if it is not, the responses of Camus's characters to the plague map nicely onto the challenge of climate change and the environmental crisis today. There are those who try to obfuscate and undersell the challenge, those who deny what is going on, those who cynically profit through manipulating fears, those who resign themselves, those who try to escape elsewhere, those who anticipate the solution arriving from elsewhere, and then those who respond in ways that might be considered heroic and rebellious even though they are quite understated. It is to them, and to the main character, Dr. Rieux, that we turn in order to think through some of the characteristics of the rebel environmental educator while also integrating our proposed negation and exaltation.

Teacher as Witness

The narrator's identity in the *The Plague* remains hidden for much of the novel, but it goes to the heart of the rebel hero's character, Dr. Rieux, to learn he is the author, and he "resolved to compile this chronicle, so that he should not be one of those who hold their peace but should bear witness in favour of those plague-stricken people; so that some memorial of the injustice and outrage done them might endure" (Camus, 1960, p. 251). Quite early in the novel Rieux, having seen the dying rats and the growing number of sick and dying patients, is called into a meeting with several doctors, politicians, and leaders of the community. The point of the meeting is to discuss strategy in response to the challenges being faced. It was at that meeting that, in the face of another doctor wanting to temper the diagnosis and several politicians wanting to understate the challenge and limit the financial expense, Rieux simply bears witness to what is happening, names it, and recommends that the city respond in the ways it must (e.g. closing its gates, setting up a quarantine system, isolating the sick, etc.).

And so one of the first acts of Camus's rebel hero Rieux was to speak truth to the politicians, other doctors, and even his patients. Unwilling to ignore its presence he named it and from then on worked in response to that named reality. Sadly, others are unwilling to respond, yet Rieux continues to name the plague, gather allies, and put the needed responses into place. Intriguingly for environmental educators, what Rieux is doing is really just following the protocols and Hippocratic Oath that exist for medical doctors in plague situations. How might that look for educators in a time of ecological crisis? By naming the environmental challenge and by using the negations and exaltations proposed above we are beginning to form the basis upon which the actions of any educator can be judged and determined with respect to the environment. For example, does the curriculum I have chosen for tomorrow appear to acknowledge, respond to, and take into account the environmental crisis? Or, did the answers I gave to my students' questions today honour their dignity and provide room for the dignity of the more-than-humans that we worked with? Or, how did I notice and respond to instances of anthropocentrism, hyper-individualism, and environmental backgrounding in the structure of our classroom, the responses of the students, and my own choice of metaphors?

Teacher as Artist

Throughout the novel, Rieux is pushed by the situation to find creative solutions to challenges as they arise. One of the clearest examples is his involvement of unexpected people. He knows that he needs people to organize the teams that are involved in gathering, transporting, and caring for the sick and he locates a previously quiet, ignored, and somewhat odd fellow to take the lead in this challenging role. In doing so, Rieux undermines the way this person has been "created" by the community and allows him to flourish in an unexpected way. For Camus, rebellion is a creative process made up of a multitude of creative acts, often unexpected, that move the community forward. And he clearly sees the focus as being on the result rather than on the artist. The point is to get a system for dealing with the ill and not about how brilliant Rieux was in choosing the person. Camus (1956) suggests, "art consists in choosing the creature in preference to his [sic] creator. But still more profoundly, it is allied to the beauty of the world or of its inhabitants against the powers of death and oblivion. It is in this way that his [sic] rebellion is creative" (p. 267). For Camus, this is a call to be in the world, in all of its beauty and complexity, with all of its denizens, and a call to assist, even if completely futile, in creating a shared mutual flourishing. In this, Camus looks to Proust, whom he admired, and to how Proust, as rebel creator, exalted life, its particularities, its uniquenesses, and its sensualities by de-centring that which, at his historical time, was the nexus around which everything else was supposed to revolve—the image in which all else was created, the metanarrative that undergirded all other stories—God. The point here is that, for today's environmental educator, there is still a creator, a metanarrative, around which everything else revolves and is understood today, and that is us: the human. In this modern, postmodern, neoliberal world we are creators and we have made the natural world—the more-than-humans—into creatures of human subjugation. We are the creator and it is this problem—this profound anthropocentrism— that the artist rebel teacher must respond to.

How might educators creatively de-centre this metanarrative and exalt mutually dignified flourishing. The environmental educator might ask— how am I inviting the local more-than-humans to be part of my teaching practice? How am I considering and creating learning environments that demonstrate that the human is not the single centre of the world? How, in honouring the chosen negation and exaltation, am I focusing on that which is created—even if it pushes me into the background? Or, how might assessment and evaluation look if we consider mutual flourishing and push individualism into the background?

Teacher as Rebel Hero

Throughout *The Plague* there are opportunities for Rieux to prioritize himself and choose to escape, stop engaging, or benefit himself in different ways, yet he does not. He is a humble, quiet hero working alongside many others in response to what appears to an overwhelming force that is killing his city. But it is clear as the novel progresses that, although the odds seem slim for survival, the only chance that exists is for everyone to find, in their own ways, something to do in response to the challenge. There is a shared foundation that supports this work and that acts as a kind of lens for the city and for each individual. It is through this core foundation that we hope this chapter has moved us a step closer to naming—exaltations, those things that we say yes to, and negations, those that we do not accept. We suggest that it might be through bearing witness and negating our own suicide as a result of our individualistic anthropocentrism, while at the same time allowing all to exercise their freedom through exalting mutual dignified flourishing, that we can, as living beings on this beautiful wild planet, survive and even thrive.

NOTES

1. Camus wrote *The Myth of Sisyphus* 5 years earlier.
2. For instance, and to push Camus's thought here, at the individual level if you cannot tell me why you are alive/what you are saying yes to then potentially your murder be allowable.
3. And murder of myriad species as well.
4. It is clear that there are cultures and peoples within the species that are more and less responsible for the destruction wrought globally. This chapter is likely aimed at those peoples that have historically taken, and continue to take, a "colonizing" position towards the more-than-human world.
5. Note: This extends Abram's concept of the more-than-human to include the uniquenesses and individualities of said members, hence the pluralized form.
6. Care is taken not to reconstruct a subject/object dualism both to remove human exceptionalism and maintain existentialist integrity.

REFERENCES

Camus, A. (1956). *The rebel* (A. Bower, Trans.). New York: Vintage Books.
Camus, A. (1960). *The plague* (S. Gilbert, Trans.). Harmondsworth: Penguin Books.
Newton, M., & Hay, P. (2007). *The forests.* Hobart: Self-published.
Sartre, J.-P. (1992). *Notebooks for an ethics* (D. Pellauer, Trans.). Chicago: University of Chicago Press.

Sean Blenkinsop is Professor of Education at Simon Fraser University and a visiting professor at SFU's Centre for Dialogue. He has a long publication record in the areas of environmental, outdoor, and experiential education as a philosopher of education. His current research interests are coming together around a discussion of significant change in education through an educational philosophy that brings together continental existentialism and eco-theory. He is also the principal investigator at two publicly funded nature-based outdoor schools, the Maple Ridge Environmental School (http://es.sd42.ca/) and the NEST program at Davis Bay Elementary (http://www.sd46.bc.ca/index.php/nature-program).

Marcus Morse grew up in Tasmania, Australia, where early journeys along the island's rivers, coastlines, and mountains, combined with the political and environmental movements of the state, inspired a long-term commitment to environmental education. Currently Senior Lecturer and Academic Programme Director of Outdoor and Environmental Education at La Trobe University, Australia, his teaching focuses on ways in which education can promote the social and cultural change required for sustainable and ecologically healthy futures. He is a lead member of the La Trobe research programme "Remaking Education" and his current interests are in the areas of environmental education philosophy, understandings of meaning-making, dialogue in education and forms of paying attention within outdoor environments. He has received a research award for his work on educational curriculum and pedagogy and is Associate Editor for the *Journal of Outdoor and Environmental Education*.

Education and the Common Good

Heila Lotz-Sisitka

Abstract The chapter responds to a recent invitation by the UNESCO to respond to the contents of their book on the purpose of education, entitled *Rethinking Education: Towards a Global Common Good?* I explore the concept of the common good (as it relates to concepts of commons and commoning activity) and what it might mean to engage with commoning as an educational activity, if the commons, as argued by Amin and Howell, is to be "released" from historical descriptions of commons and commoning activity, to embrace a futures orientation. Drawing on critical realism and decolonization theory, as well as experience of working with expansive social learning, I propose that an educational theory grounded in a concept of emergence is needed in such a context.

Keywords Education · Transformation · Transformative learning · Social learning · Humanizing · Commons · Commoning · Common good · Emergence

H. Lotz-Sisitka (✉)
Environmental Learning Research Centre, Rhodes University, Grahamstown, South Africa
e-mail: H.Lotz-Sisitka@ru.ac.za

© The Author(s) 2017
B. Jickling, S. Sterling (eds.), *Post-Sustainability and Environmental Education*, Palgrave Studies in Education and the Environment,
DOI 10.1007/978-3-319-51322-5_5

The chapter responds to a recent invitation by UNESCO's panel of advisors to step into dialogue with the contents put forward in a booklet released by UNESCO entitled *Rethinking Education: Towards a Global Common Good?* (UNESCO, 2015). This document is potentially significant, as it provides new orientation to educators around the world, replacing the influential Delors report (1996) that many governments used to frame the purposes of education. The new UNESCO (2015) booklet proposes that in rethinking the purpose of education, there is a need to link between objectives of sustainable development and the concept of a global common good, and it suggests that the educative modality for this ought to be humanistic pedagogy. In doing so, it argues that education in itself is, and must be, a global common good. In addition to this affirmation, UNESCO suggests a new purpose for education by suggesting that, "Education must be about learning to live on a planet under pressure" (2015, p. 3). UNESCO thus suggests a reorientation of the purpose of education, and goes further to say that education, while traditionally oriented towards enculturation and adaptation, can also be oriented towards transformation. While interesting for education, especially as the document provides an invitation to educators to consider its propositions for reorienting the future purpose of education, the 2015 UNESCO document lacks in-depth guidance on what exactly is meant by the relevance of education as a common good, a topic that I consider below.

THE COMMONS, THE COMMON GOOD, AND COMMONING AS ACTIVITY

In 2015, shortly before he passed away, Roy Bhaskar's last recorded words, shared posthumously in a book on his theory of education, were:

> Our starting point is to remember that we are natural beings and we depend on the sun.... [An] anti-naturalist perspective is an implicit tendency in a lot of Western thought [and by implication our education systems and modern knowledge]. Society arises out of nature, and the more we differentiate ourselves from it, the more problems we have. (Bhaskar, in Scott, 2015, p. 41)

Bhaskar's critical realist theory of transformative praxis (2008) and his associated theory of education via ethics and emergence in open systems are premised on a notion of common good which proposes that the flourishing of one is related to the flourishing of all (Bhaskar, in Scott,

2015). Bhaskar's work explicitly brings the tensed relation between situation, solidarity, and freedom—which lies at the heart of many debates about the commons, commoning as activity, and the common good—to the fore. The commons refers here to those spaces, resources, ways of being, and systems (e.g. the earth systems) that are shared by all. Commoning as activity involves individual and collective actions to take care of shared resources, ways of being, and systems in the interests of social justice and ecological care. One could surmise that it also lies at the heart of UNESCO's attempt to define a purpose of education at the intersection of sustainable development, the global common good, and humanistic pedagogy (UNESCO, 2015). However, for this intersectional discourse to have meaning in the life of the world's majority people, I propose that a more careful analysis of the common good and education is needed from a sustainability perspective.

Many environmental and social justice movements have, for some time now, shown a strong commitment to the common good and the commons (Slater, 2004; Ostrom, 2010; Pithouse, 2014; Nixon, 2011; Martinez-Alier et al., 2014). In the environmental arena, Vandana Shiva (1992, 2005) is one of the more explicit champions of a political position that seeks to reclaim the commons from neoliberal forces. In developing her position, Shiva supports a strong notion of "Earth Democracy," which she argues is intimately tied to democratization and de-commodification. Her work focuses on both the material and immaterial commons, such as enclosure and commodification of water, seeds, and land, as well as biopiracy and the patenting of traditional ecological knowledge. In decolonization commons work, Dussel (1998), Shiva (1992, 2005), De Sousa Santos (2014), and others involved in environmental justice work (Martinez-Alier et al., 2014) suggest that environmental justice and sustainability will only have meaning in relation to the commons if the commons are to be interpreted in relation to the wider common good— that is, if it were to include the majority of the world's people, as well as the earth's system and the more-than-human world. Amin and Howell (2016) suggest that it is perhaps because of the massive enclosure, encroachment, and loss of the commons for the world's majority that the "consciousness of a collective common life" endures and that this is intimately linked to "the planetary precarity that affects us all, and all forms of life" (241[1]). The fate of the Arctic, they suggest, is iconic in this regard.

Linebaugh (2008, 2014) critiques tendencies to privatize land and the commons as a long-term structural feature of modernity. He and environmental justice researchers critique the work of Hardin's (1968) thesis of the "tragedy of the commons" for perpetuating this orientation. The critical historical work of Linebaugh (2008, 2014), and the empirical work of Nobel Laureate Elinor Ostrom (1990, 2010) has shown, contrary to the Hardin thesis, that people *do* work together in collectives to manage common property for common good gains, and that they have done so under extremely precarious conditions. However, their work has primarily focussed on historical and contemporary enclosures of the material and biophysical commons (e.g. land, forests, waters, and fisheries). Gibson-Graham, Cameron, and Healy (2016) suggest that Hardin's (1968) work might be reinvoked with more care given to the concept of "unmanaged" that was, according to him, neglected in readings of his work.

Today's environmental crisis creates new conditions for rethinking commons theory and practice, since the commons that is being referred to crosses boundaries and nation state borders and is therefore both difficult to manage and thus "unmanaged" in a traditional sense (Gibson-Graham et al. 2016). Therefore, traditional forms of management (e.g. local action and policy at the nation state level) may also be inadequate in and of themselves, in the light of the form of commons engagement that might be required in the face of climate change and associated intersectional concerns. It is here that educational engagement with the commons and commoning activity has potential. Gibson-Graham (1996) suggests a new form of politics around commoning in the Anthropocene that is not bound by what they call a "capitalocentric" framing. Similarly, Amin and Howell (2016) suggest "Practices of commoning need to be extended to a more-than-human community as well as to a more-than-capitalist one" (387[2]).

Associated with this is the work of critical thinkers who suggest that the scale of the climate crisis requires a different way of thinking about humans and human activity (Chakrabarty, 2009; Plumwood, 2007) involving a "deep history approach" in which humans see themselves not just as one species amongst a multispecies community of life, but rather a species whose existence depends on other species (Gibson-Graham et al., 2016, 5240[3]; Dussel, 1998, see also Bhaskar, 2015 above). In this context, Amin and Howell's (2016) recent book entitled *Releasing the Commons: Rethinking the Futures of the Commons* suggests that there is a need to move beyond seeing the commons in the past tense. They argue that the commons should also be seen not "as an entity passed over from the public into the private" and that there is need to reimagine the commons "as a *process*, a *contest of force*, a reconstitution, and *a site of convening practices*"

(p. 3, my emphasis). Linebaugh (2008) introduces the concept of "commoning," which turns the noun into a verb, and suggests that there are no commons and thus no orienting to the common good without commoning. He suggests:

> To speak of the commons as if it were a natural resource is misleading at best and dangerous at worst—the commons is an activity and, if anything, it expresses relationships in society that are inseparable from relations to nature. It might be better to keep the word as a verb, an activity, rather than as a noun, a substantive. (p. 279)

This suggests a need for bringing new practices of "being in common" into existence, that is, *commoning activity*. Despite an almost overwhelming play of forces in the modern day that promote hyper-privatization and commodification of life, the possibility exists of the commons being reclaimed within a "contested and dynamic domain of collective existence, with the balance delicately poised between rapacious demands of political economy and the promise of social innovation" (Amin & Howell, 2016, p. 192[4]). Amin and Howell propose commoning activities such as establishing social solidarity networks, creating community economies, and embracing the knowledge commons amongst others. Rather than to promote nostalgia or fantasy in discourses of the commons and common good, there is need to focus our attention on commons building or "commoning activity" (Amin & Howell, 2016; Linebaugh, 2008, 2014). What neither Amin and Howell (2016) or Linebaugh (2014) do, however, is develop a theory of how education and learning play a role in the building of commoning activity, although there is recognition that learning is an important process associated with commoning activity.

Commoning Activities and Expansive Learning as Acts of Humanizing Education

One of the proposals in the UNESCO (2015) booklet on rethinking education is that education for the common good must be achieved via a humanizing approach which incorporates ethics and values and is "based on respect for life and human dignity, equal rights, social justice, cultural diversity, international solidarity, and shared responsibility for a sustainable future" (p. 10) which UNESCO sees as "the fundamentals of our common humanity" (p. 10). Here, I consider how engaging in expansive learning to shape and grow commoning activity may offer possibilities for a humanizing approach that also takes the more-than-human into account. Bhaskar (in Scott 2015) suggests developing a constellational relational understanding of material reality and how this relates to human agency, knowledge, learning, and change in ways that take the more-than-human into account, ethically. His theory of education, like that proposed by the commitment to emergent processes in expansive learning research, (Engeström, 1987, 2007; Engeström & Sannino, 2010, 2016) affirms the principle of emergence in open systems which is necessary for the building of commoning activity in new conditions (De Angelis, 2006).

In our research programme in Africa, we have been engaging in small-scale experiments to develop education and learning processes that provide insight into the possibilities for this way of thinking of education and the emergence of commoning activity. In this work, the focus has been on expanding social learning potential and mobilizing intentional agency for collective commoning activity. Examples of such research include: expanding learning-centred possibilities for co-management of fisheries resources in Malawi (Kachilonda, 2015, see below); expanding learning-centred possibilities for rainwater harvesting practice and small-scale sustainable agriculture amongst rural poor communities working on communal food gardening, and their local economic development centres and agricultural colleges in South Africa, Lesotho, and Zimbabwe (Pesanayi, 2016; Mukute, 2010; Mukute & Lotz-Sisitka, 2012; Mukute, 2016).

In engaging in such learning-centred experiments in open systems of commoning activity, we are of course reminded that in our educational activity we should also not overplay a hope-filled, affirmative stance (Amin & Howell, 2016), but neither should we give up the possibility of hope and social change. The examples cited above all show real gains in terms of supporting collective transformative agency via learning processes, especially insofar as the expansive learning process (Engeström & Sannino, 2010, 2016) enables the envisioning of alternative futures and possibilities. They also show how small-scale commoning practices and activities that will be needed in making practical sense of the concept of education as a common good can be realized, even in contexts where the odds are stacked against those attempting such commoning activities.

The spaces for commoning activity that have formed the focus of our small-scale research experiments offer a potentially interesting option for rethinking education. The work offers new forms of both *being* and *becoming* via education. It shifts the focus on education oriented to enculturation only, to education oriented towards aspiration, or what Appadurai (2014) has named "culture as future," and a form of "learning activism" (Choudry, 2015).

Solnit (2014) argues that commoning can become a way of being in the world. As argued in the above, Bhaskar and Engeström's theories of education provide a means of us also thinking that commoning can be a way of not only being in the world but also of becoming, that can be realized in processes of transformative praxis in collective learning settings, similar also to the generative impulses in the work of Freire (1970). As such, these theories of education may also provide a means of rethinking education, not as access to forms of existant "powerful" knowledge (Wheelahan, 2010), but rather as means of engaging the social-ethical dynamics associated with being and becoming in the world, in which knowledge is important, but not the only defining factor (Zipin, Fataar & Brennan, 2015). Zipin et al. (2015) eloquently argue for a return to ethics as a foundation of knowledge engagements in education and peda-gogy. They also argue for bringing social-subjectivity to the fore in educa-tional decision-making. Bhaskar (in Scott 2015) proposes an onto-epistemic foundation for education and learning, in which knowledge is important, but where knowledge should not be reified and disembedded from the constellational socio-material processes of being and becoming. Bhaskar suggests that such a form of learning involves new kinds of "working at it" at different levels of being and becoming (Price & Lotz-Sisitka, 2016). Scoping the focus of such education and learning he says:

> Yes, we have to work at it. The work, and this is crucial, the work has to be done on all four planes of social being. These are material transactions with nature, social interactions between people, social structure, and the stratifi-cation of the embodied personality. (Bhaskar, in Scott, 2015, p. 18)

Importantly also, in providing perspective on my arguments above about the potential role of education in commons-building, is the insight from Amin and Howell (2016) that "solidarity does not depend on social praxis alone." There is also need for solidarity-oriented social praxis to begin to shift laws and policies, monopolies, unregulated trade, over-privatization tendencies, property exclusions, and the like. However, this in turn requires giving attention to the relationality in social transformation pro-cesses. As Archer (2015, p. 1) argues, "any social formation has a particular *relational organisation* between its parts," which includes relations between structural, cultural, and agentive emergent properties, which she also explains are generative mechanisms that shape social transformations. In explaining transformation towards the common good, there is therefore need to give adequate attention to the nature of relational organization. "In the social order, generative mechanisms always exist in the plural and are thus in interplay with one another, conjointly producing what actually happens in the world" (Archer, 2015, p. 3). Such generative mechanisms are activity dependent which means that when we refer to commons, we ought to refer to directly to the "specific social process(es) in question" (Archer, 2015, p. 4). That is, we ought to refer to the specific commoning activity (e.g. co-operative organic farming) and the generative mechanisms that shape the activity in any particular context.

Archer suggests that this requires social theorists [and educators] who are engaged in transformative praxis processes such as those proposed above for education, to recognize multiple sources of determination, and this also makes deterministic accounts impossible. In this way, our explanation of commoning activities and expansive learning processes can be realist, and not dependent on empiricism or universal actualism. As Bhaskar says, "the acts of creation exist in themselves as emergent and novel discoveries about the world" (in Scott, 2015, p. 32). Such creative discoveries, are however, also related to knowledge, both knowledge that pre-exists us in such contexts, but also knowledge that we are trying to develop.

I offer one brief account of a realist explanation of an expansive learning commoning activity from our research programme. In Kachilonda's Ph.D. study, fishers engaged collaboratively with government officials, college lecturers and students, researchers, and local traditional leaders to consider how to better co-manage the Lake Malawi fishery. Kachilonda states:

> The historically constituted command-control approach to fisheries management generally led to the creation of distrust and disloyalty among communities and increased levels of conflict amongst communities and government officials. This was the context then, with fish stocks declining and local economies crumbling, when the Malawi government realised something needed to be done, and introduced policies of co-management. (2015, p. 8)

These policies were, however, poorly implemented. Through an interactive and expansive social learning process in his research context, the multi-actor groups were able to identify contradictions that were hampering the collective co-management of the fishery, and to seek forms of action and relationality that facilitated changes in practice towards improved co-management of the fishery. For example:

> Getting less fish catches means starving our families because fishing is the only source of income in this area. The little income we get from fishing is used to support our families in the whole of Lake Malombe. (Focus group participant, in Kachilonda, 2015, p. 144)

> Sometimes we as Beach Village Committee members are not satisfied with the chief's judgement and when that happens we inform our extension officers to help and talk to the chief. (Interview with BVC committee member; in Kachilonda, 2015, p. 146)

This allowed for multilevelled forms of transformative learning and social praxis, that built social movement along the chain of human activity required for substantive societal transformations to occur in co-management practices in the Lake Malawi context. This social movement is reflected in the following statements:

> We used to experience some problems when the arrangement of co-management started because we were not trained on how to work together. After several sessions we saw that our relationships with fellow fishers and those from government were getting better. Things are getting clearer and every time we meet we learn new things and everyone strives to improve practices. (Interview with individual fisher, in Kachilonda, 2015 p. 146)

> The coming of co-management in Lake Malombe has changed the way we used to interact when the government was in control of the fisheries activities. As people who have similar interests we are able to sit, discuss and agree on certain issues. We can say with co-management, fishers are able to discuss and agree on better ways of managing the fisheries resources and people share past experiences and map for the future as resource users. (Interview with individual fisher, in Kachilonda, 2015, p. 146)

This multilevelled form of expansive social learning across the diverse activity systems engaged in co-management became possible through engaging diverse actors in collective learning engagements over a period of time, in which they all reflected on, and generated new processes for collectively ensuring better co-management of the commons resource, the Lake Malawi fishery. In so doing, there were a range of generative mechanisms shaping the possibilities for their learning and agency. These included long histories of poverty, cultural diversity, bifurcation of governance that emerged via colonial state formation, and inadequate knowledge of biodiversity and ecosystem collapse. An example of how cultural diversity and migration patterns, as generative mechanisms, shaped the commoning activity is described in a focus group interview with fishers:

> We have seen over the years that we are no longer people belonging to one group. We now have people from the north who have come with different fishing practices which have also brought in different cultures and understanding in the fishery. If we were to take xx's example, he only accepts those who comply with him and those who do not are not allowed to do fishing in the area. We can gain a lot from the knowledge they bring but we also need to respect our systems. Our groups have rules to follow and I think those should be followed. (Focus group interview with fishers, in Kachilonda, 2015, p. 152)

Taking account of the generative mechanisms within a generative complexes frame (after Bhaskar) in this research was essential for heeding Amin and Howell's (2016) caution of nostalgia or fantasy in pursuit of the common good.

TOWARDS A CONCLUSION

Twentieth-century education had a large role to play in creating a chain of transformations in human activity oriented towards industrialization, commodification, and the current dominance and exclusionary nature of the neoliberal economic trajectory. Is it possible that a form of twenty-first-century education can be born via expansive learning opportunities that are oriented towards the common good? And that these can influence complex chains of human activities (Engeström & Sannino, 2016), not in the direction of a few having all as in twentieth-century education, but in the direction of more being shared more equitably and sustainably by all? This, in my view, is the critical question to be considered in relation to UNESCO's invitation to step into dialogue with their proposal that education ought to be a common good, oriented towards the common good.

This requires educators to engage critically and imaginatively with the proposed intersectionality of sustainability, the global common good, the humanistic orientation put forward by UNESCO in their 2015 direction setting document, and the more-than-human world. But this also needs to be actualized in an emergent process of expansive learning. This offers real-world potential for dialogue with the UNESCO paper and its propositions.

NOTES

1. Location 241 of 7081 in e-book version
2. Location 387 of 7081 in e-book version
3. Location 5240 of 7081 in the e-book version
4. Location 192 of 7081 in e-book version

REFERENCES

Amin, A., & Howell, P. (2016). *Releasing the commons: Rethinking the futures of the commons.* London: Routledge.

Appadurai, A. (2014). *The future as cultural fact: Essays on the global condition.* London: Verso.

Archer, M. (2015). Introduction: Other conceptions of generative mechanisms. In M. Archer (Ed.), *Generative mechanisms transforming the social order* (pp. 1–26). Switzerland: Springer International Publishing.

Bhaskar, R. (2008). *Dialectic: The pulse of freedom.* New York: Verso.

Chakrabarty, D. (2009). The climate of history: Four theses. *Critical Inquiry, 35* (Winter), 197–222.

Choudry, A. (2015). *Learning activism: The intellectual life of contemporary social movements.* Toronto: University of Toronto Press.

De Angelis, M. (2006). Editorial. *The Commoner, 11*(Spring/Summer), 1–3. Retrieved from http://www.commoner.org.uk

De Sousa Santos, B. (2009). *Epistemilogicas do Sul. Coimbra: Almendina [Translated into English as Epistemologies of the South: Justice against epistemicide].* [2014]. London: Verso.

Delors, J. et al. (1996). *Learning: The treasure within.* Paris: UNESCO.

Dussel, E. (1998). Beyond Eurocentrism: The world-system and the limits of modernity. In F. Jameson & M. Myoshi (Eds.), *The cultures of globalization* (pp. 3–31). Durham, NC: Duke University Press.

Engeström, Y. (1987). *Learning by expanding: An activity theoretical approach to developmental research.* Helsinki: Orienta-Konsultit.

Engeström, Y. (2007). Enriching the theory of expansive learning: Lessons from journeys towards co-configuration. *Mind, Culture and Activity, 14* (1–2), 23–39.

Engeström, Y., & Sannino, A. (2010). Studies of expansive learning: Foundations, findings and future challenges. *Educational Research Review*, doi:10.1016/j.edurv.200212.002

Engeström, Y., & Sannino, A. (2016). Expansive learning on the move: Insights from ongoing research. *Infancia y Aprendizaje /Journal for the Study of Education and Development, 39*(3), 401–435. Retrieved from http://dx.doi.org/10.1080/02103702.2016.1189119

Freire, P. (1970). *Pedagogy of the oppressed.* (Myra Bergman Ramos, Trans.). New York: Continuum.

Gibson-Graham, J. K. (1996). *The end of capitalism (As we knew it): A feminist critique of political economy.* Oxford: Blackwell.

Gibson-Graham, J. K., Cameron, J., & Healy, S. (2016). Commoning as post-capitalist politics. In A. Amin & P. Howell (Eds.), *Releasing the commons: Rethinking the futures of the commons* (pp. 192–212). London: Routledge.

Hardin, G. (1968). The tragedy of the commons. *Science, 162*(3859), 2343–1248.

Kachilonda, D. (2015). *Investigating and expanding learning in co-management of fisheries resources to inform extension training.* (Unpublished PhD thesis). Rhodes University, Grahamstown.

Linebaugh, P. (2008). *The Magna Carta manifesto: Liberties and commons for all.* California: University of California Press.

Linebaugh, P. (2014). *Stop thief! The commons, enclosures and resistance.* Oakland, CA: PM Press.

Martinez-Alier, J., Anguelovski, I., Bond, P., Del Bene, D., Demaria, F., Gerber, J.-F., Yánez, I. (2014). Between activism and science: Grassroots concepts for sustainability coined by Environmental Justice Organizations. *Journal of Political Ecology, 21*, 19–60.

Mukute, M. (2010). *Exploring and expanding learning in sustainable agricultural practices in Southern Africa.* (Unpublished draft PhD). Rhodes University, Grahamstown, South Africa.

Mukute, M. (2016). Dialectical critical realism and cultural historical activity theory (CHAT): Exploring and expanding learning processes in sustainable agriculture workplace contexts. In L. Price & H. B. Lotz-Sisitka (Eds.), *Critical realism, environmental oearning and social-ecological change* (pp. 190–211). London: Routledge.

Mukute, M., & Lotz-Sisitka, H. (2012). Working with cultural-historical activity theory and critical realism to investigate and expand farmer learning in Southern Africa. *Mind, Culture, and Activity, 19*(4), 342–367.

Nixon, R. (2011). *Slow violence and the environmentalism of the poor.* Cambridge, MA: Harvard University Press.

Ostrom, E. (1990). *Governing the commons: The evolution of institutions for collective action*. Cambridge, MA: Cambridge University Press.

Ostrom, E. (2010). Beyond markets and states: Polycentic governance of complex economic systems. *American Economic Review*, *100*(3), 641–672. doi:10.1257/aer.100.3.641

Pesanayi, T. (2016). Exploring contradictions and absences in mobilizing "learning as process" for sustainable agricultural practices. In L. Price & H. B. Lotz-Sisitka (Eds.), *Critical realism, environmental learning and social-ecological change* (pp. 230–253). London: Routledge.

Pithouse, R. (2014). An urban commons? Notes from South Africa. *Community Development Journal*, *49*(suppl 1), i31–i43. doi:10.1093/cdj/bsu013

Plumwood, V. (2007, (August–September)). A review of Deborah Bird Rose's *report from a wild country: Ethics of decolonisation*. *Australian Humanities Review*, *42*, 1–4.

Price, L., & Lotz-Sisitka, H. (Eds.), (2016). *Critical realism, environmental learning and social-ecological change*. London: Routledge.

Scott, D., with Bhaskar, R. (2015). *Roy Bhaska:. A theory of education*. Switzerland: Springer International Publishing.

Shiva, V. (1992). The seed and the earth: Women, ecology and biotechnology. *The Ecologist*, *22*(1), 4–8.

Shiva, V. (2005). *Earth democracy: Justice, sustainability, and peace*. Cambridge, MA: South End Press.

Slater, D. (2004). *Geopolitics and the post-colonial: Rethinking North-South relations*. Malden and Oxford: Blackwell Publishing.

Solnit, R. (2014). *The encyclopaedia for trouble and spaciousness*. San Antonio, TX: Trinity University Press.

UNESCO. (2015). *Rethinking education: Towards a global common good?* Paris: UNESCO. Retrieved from http://unesdoc.unesco.org/images/0023/002325/232555e.pdf

Wheelahan, L. (2010). Competency-based training, powerful knowledge and the working class. In K. Maton & R. Moore (Eds.), *Social realism, knowledge and the sociology of education: Coalitions of the mind* (pp. 93–109). New York and London: Continuum.

Zipin, L., Fataar, A., & Brennan, M. (2015). Can social realism do social justice? Debating the warrants for curriculum knowledge selection. *Education as Change*, *19*(2), 9–36. doi:10.1080/16823206.2015.1085610

Heila Lotz-Sisitka works in the Environmental Learning Research Centre at Rhodes University in South Africa, where she holds a South African National Research Foundation Chair focusing on transformative social learning systems. Her current research focuses on critical research methodologies and the relationship between learning, agency, and social-ecological and education system change for the common good. She is Principal Investigator of the International Social Science Council research programme on *Transgressive Social Learning for Social-Ecological Sustainability in Times of Climate Change* which involves establishment of co-engaged research in a nine-country Transformative Knowledge Network across four continents. She is co-editor of the recent book *Critical Realism, Learning and Social-ecological Change*, published by Routledge in 2016.

Experience and Relation

Sustainability and Human Being: Towards the Hidden Centre of Authentic Education

Michael Bonnett

Abstract It is argued that while the idea of sustainable development fails adequately to provide a way forward for addressing our current environmental crisis, the idea of sustainability contains the germ of an understanding of the character of human consciousness that places our relationship with nature at the heart of both human being and authentic education. A phenomenological approach is developed to explore some key aspects of our experience of nature—particularly its "otherness" and intrinsic normativity. The potential of the mutually sustaining relationship with nature that emerges for shaping our outlook on the world, and revealing the subverting effects of the scientism that is taken to permeate contemporary Western-style culture, are discussed. Some broad implications for a re-envisioned education are sketched.

Keywords Authentic education · Human being · Intentionality · Nature · Scientism · Sustainability

M. Bonnett (✉)
Independent Scholar, UK Universities of Cambridge, London, Bath, England
e-mail: mrb25@cam.ac.uk

© The Author(s) 2017
B. Jickling, S. Sterling (eds.), *Post-Sustainability and Environmental Education*, Palgrave Studies in Education and the Environment,
DOI 10.1007/978-3-319-51322-5_6

The threats posed by anthropogenic global environmental degradation (including climate change) have led many to seek to incorporate some version of education for sustainable development into the curriculum. Frequently, this has taken the form essentially of some sort of bolt-on addition to a superordinate pre-existing educational structure whose basic motivations were elaborated without any reference to ideas of sustainability. This chapter argues that there is a key sense of sustainability that, far from being something to be conceived as a contingent addition to the curriculum, lies at the very heart of the meaning of authentic education, and that requires us to reappraise what largely today has come to count as education in the West.

THE IDEA OF SUSTAINABLE DEVELOPMENT

As an instrument of change, the idea of sustainable development has failed. Although still prominent in much of the rhetoric concerning environmental issues, at a cultural level it fails sufficiently to motivate us. Despite all the scientific evidence, as a culture we continue in practices that seriously degrade nature and heat the Earth. For many in the West, daily life lacks any feeling of oneness with the natural world in which our being is ultimately embedded. Indeed, in many respects, we are still behaving like irresponsible adolescents. This empirical state of affairs is entirely consistent with the criticisms that have been raised against taking sustainable development as an orienting idea for environmental education. While I will not rehearse these criticisms here (but see, e.g., Jickling, 1992; Bonnett, 1999), in the context of the discussion to be developed in this chapter it is worth briefly revisiting one that has frequently arisen: the tension that exists between the idea of sustaining and the idea of development. With regard to thinking about the natural world, "sustaining" has strong conservationist strands that embody ideas of respecting and nurturing things as they are in their own nature. "Development," on the other hand—and in the context of the problems that the idea of sustainable development was intended to address— emphasizes the idea of planned change: an anthropocentric ordering of things so as to produce more or better than what currently exists. In practice, in the context of Western-style culture it has become closely affiliated with materialistic economic growth that in turn involves ever more extensive demands upon, and manipulation of, the natural world.

This is not of course to deny that development is inherent in the natural world, far from it. Rather it is to recognize that the kind of development involved in dominant interpretations of sustainable development is of a radically different sort to that found in nature. The difference can be characterized by saying that with the former, development proceeds in accordance with an external norm that bears little or no relationship to the internal norm of the thing's own development. An example of this would be plant and animal breeding programmes that seek to "improve" upon nature in the sense of engineering outputs that better meet current human desires, but that no longer relate to their flourishing in their natural environment.[1] Contrary to this inclination, at the heart of sustainability lies the idea of allowing things in nature to be themselves, rather than to reshape and redirect them in accordance with criteria that do not emanate from themselves.

But wait! Surely, this way of putting things cannot be right. With the exception of pure wilderness—if any still exists—there is no pristine nature to develop according to purely internal norms. There is no nature unaffected by, unmodified by, human activity or its consequences. As some put it, there exists now only "second nature"—nature so-modified. Furthermore, if we seek insight into what is happening on a global scale frequently we need to make reference to an abstract "third nature" that makes its appearance through complex computer modelling. Increasingly, for many, "first nature" is becoming a dim recollection or an irrelevant myth: in so many ways (physically and conceptually), it is claimed that nature is best regarded as a social construction rather than some transcendent source of being and authority.

This sentiment is reflected in Ursula Heise's summary of how nature frequently is regarded in academia:

> More broadly, the basic goal of work in cultural studies for the last twenty years has been to analyze and, in most cases, to dismantle appeals to "the natural" or the "biological" by showing their groundedness in cultural practices rather than facts of nature. The thrust of this work, therefore, invariably leads to skepticism about the possibility of returning to nature as such, or of the possibility of places defined in terms of their natural characteristics that humans should relate to. (Heise, 2008, p. 46)

In the light of this trend the first task must be to elucidate a conception of nature that not only can rebut such attempts to erode its significance, but that also can help to reveal the motivations that energize them. I turn to our immediate experience of nature.

A Phenomenology of Nature

The key general characteristic of our direct experience of nature is that it presents itself as independent of our will. Of course we can affect nature in all sorts of ways, but in all our interactions with it there remains something beyond us, a given. However far the frontiers of our manipulations reach there remains the encounter with something already there—other, not of our authorship. In this sense quintessentially nature is experienced as *self-arising* (Bonnett, 2004).

To claim this self-arising character as essential to our understanding of nature is not to deny that frequently—perhaps for the most part—how nature reveals itself reflects the purposes that we pursue with regard to it and the concepts that we employ to articulate it. But these purposes and concepts can never fully determine what lies before us. Indeed, on occasion our ongoing application of these purposes and concepts to our experience of nature can be rebutted: for example, our attention can suddenly be commanded by something quite other with respect to our current preoccupations. Perhaps, while on a meditative walk our tranquillity is interrupted by the growing awareness of the high-energy hum of a swarm of angry bees.

More generally, there is a complex and intimate interplay between cultural motives and artefacts and the otherness of nature that constitutes world-formation: the design of an implement such as a spade both is shaped by the soil it cleaves and in cleaving brings to presence the resistance, texture, odour, and shy lustre of the clay. Similarly, the ever evolving and historically grown overarching form of sensibility through which we engage with the world has arisen in fundamental respects in response to nature as the self-arising. Indeed, through the millennia engagement with the otherness of nature has not only shaped our senses, but in large part is their *raison d'etre*. Such engagement with the otherness of nature is also implicit in the logic of notions fundamental to how we think, such as perception, understanding, and description. These notions all presuppose an independent reality to *be* perceived, understood, or described. In this sense, nature as the self-arising is ineluctably embedded in human being.

As quintessentially other-than-human, things in nature have aspects that always lie beyond us, withdrawn, as yet (and perhaps never) to be revealed. This remains true no matter how developed our scientific understanding becomes. Indeed, preoccupation with scientific observation, categorizing, and explanation can lead to an attenuation of a more direct and intimate sense of the being of things in nature. For example, the quiet mystery of the sheer existence of some wayside flower and its subtle changes of hue and aspect in the play of sunlight can easily become occluded as we turn to identifying it according to generic characters listed in a flora database. When experienced in their native occurring, things in nature are epistemologically mysterious and retain the ability to offer invitations to participate in their being in unique and never wholly predictable ways.

This brings me to another important consideration: things in nature exist always in reciprocal relationship. Scientific ecology construes this relationship in terms of causal/probabilistic law-governed biophysical interdependencies. In contrast, a phenomenological perspective that attends to the very occurring of things—the character of their living presence—sees how things in nature occur in a mutual sustaining relationship. There is an important sense in which things in nature participate in each other and thereby in a place-making (Bonnett, 2009). Take, for example, the living presence of a beech tree encountered in some woodland dell in spring. The manner in which this tree occurs for us arises in interplay between the tree and its neighbours, the play of sunlight on its shimmering leaves, the rustle of birds flitting through its canopy, the aggregations of moss and lichen on its limbs, the enveloping odours of growth and decay. By participating in this interplay the tree both contributes to sustaining the place in which we came across it, and is sustained by it. Transplanted to, say, a city mall, its being is transformed from a sheltering presence, to, perhaps, a silhouette on neon.

Implicit here, is the way that the natural world is vibrant with anticipation. It reverberates in the predator seeking out its prey, leaves unfurling for the light of day, pale roots drawing towards moisture in dark places. If we are attentive, our being becomes infused with this interplay of anticipations. Without it, we enter ontological free fall: our lives untouched and unsustained by a world that we pass through but do not inhabit. There is no sense of oneness with the sway of natural occurrences in which our authentic being is embedded—as when, under the influence of scientism, we can be persuaded that to recognize nature's transcendent inviting otherness is to indulge a frothy fiction.

Here we are brought up against ways in which things in nature possess their own integrity. Our sense of this can be evoked if perhaps we were to find the dell strewn with the remnants of fly-tipping, or the beech tree wantonly vandalized. More positively, in experiencing myriad interplays, harmonies and contrasts, subtle adaptions and accommodations, we might be struck by a sense of the rightness of the dell. Here, things occur in such a way that how they *are* communicates that this is how they *ought* to be (Bonnett, 2012). In this sense they are normative and possess intrinsic value. Though they are other, in many cases we can have some sense of what counts as their fulfilment.

It is important to acknowledge here that we, too, are sustained by our participation in such place-making. Witnessing such a scene, our being can be enlivened and enlarged, our senses awakened and refreshed, our bodies resonating with what lies before us. The being of the dell becomes our being.

SUSTAINABILITY AND AUTHENTIC HUMAN BEING

In the previous section, I briefly explored an alternative to the prominent scientific interpretation of ideas of interrelatedness and sustainability that contextualizes them in systems of law-governed interactions and causal interdependencies in biophysical nature. I will now argue that there is another very important dimension of relatedness lying in the wings: the sense in which the conscious self is always self-in-relationship.

Historically, this was powerfully articulated in the intentionality thesis that Franz Brentano retrieved from the medieval schoolmen and reintroduced into modern philosophy in the late nineteenth century. The notion of intentionality was employed by Brentano in order to distinguish between physical and psychical (psychological) phenomena. In modified form this distinction became central to Edmund Husserl's development of phenomenology. In *Psychology from an Empirical Standpoint* (1995 edition) Brentano interpreted the idea of intentionality as "relationship to a content, the tendency towards an object" that is immanent—that is, contained within consciousness as some sort of mental entity. While agreeing that a central feature of consciousness is that it is "minded" in the sense of being directed towards an object, Husserl seminally transformed the argument. In his *Logical Investigations* (2001 edition, pp. 126–127), he pointed out that this intentional object should in no way be regarded as some sort of psychological entity (such as an image, or idea) found *within* consciousness. To the contrary, it is transcendent. Its existence lies beyond any individual consciousness.

For example, when we desire or seek something—say, a new pair of shoes—consciousness is directed towards an actual object, not some image or idea that consciousness already has within it. The fact that in this key sense intentional objects are necessarily transcendent to consciousness is not to say that they necessarily physically exist (as when we think, say, of a unicorn). Rather consciousness is simply "minded" in the sense of being directed at something beyond itself. Furthermore, as Heidegger (1972) makes clear, in our experience this object exists always and already within a world. And this is a world that it shares with us. The shoes are out there as part of the world that we inhabit, sitting on some shelf in a store waiting to be purchased. Even unicorns as ideas, as mythical creatures, are experienced as existing in a world—both their own, and ours. They never occur as completely isolated entities, inhabiting nowhere.

Although he makes no explicit reference to any intentionality thesis, it seems to me that strong resonances with this thesis can be detected in Bertrand Russell's introductory text *The Problems of Philosophy* (1959 edition). In the concluding chapter "The Value of Philosophy," he suggests that true knowledge is a union of the "Self" with the "not-Self." Here knowledge does not occur through study that "wishes in advance that its objects should have this or that character, but adapts the Self to the characters which it finds in its objects. . . . In contemplation . . . we start from the not-Self, and through its greatness the boundaries of Self are enlarged; through the infinity of the universe the mind which contemplates it achieves some share in infinity" (p. 92).

This conditioning of consciousness by its objects is a key implication of the intentionality thesis for education. In combination with the analysis of sustainability developed above, the thesis provokes consideration of an important sense in which sustainability lies at the heart of human consciousness and its fulfilment. If the essence of consciousness is that it is ecstatic in the sense of constantly standing out towards things beyond itself, if the revelation of otherness is the *raison d'etre* of the senses, then the more receptive consciousness is—the more fully it participates in the manifold being of the things it encounters—the richer is its own life. And in being the place where things can come to presence—show up in their significances—consciousness lets them be. It sustains them as living presence. As always already in relationship with a transcendent world, consciousness is inherently environmental. It sustains that world and is sustained by it. Its essence lies in participating in a mutual sustaining. And this world with which consciousness is primordially engaged (including, especially, the natural world) is far removed from that fabricated and orchestrated by classical science that so often is allowed the role of arbiter of the real. Rather than a world of scientific objectivity, it is a world suffused with intrinsic value (including moral and aesthetic), normativity, agency, meaning, and feeling.

Of course, reference to things themselves here is not to be conflated with Kantian things-in-themselves. The appearing of things will always be modulated by the ways in which consciousness relates to its objects, for example, wishfully, believingly, instrumentally, etc. And all this will occur within an overarching form of sensibility with its current but always evolving and interplaying sensory, affective, and cognitive structures. Only through this modulation can significances show up. The appearing of things is thus always human related, but it need not be human centred. Indeed, insofar as we wish to live in truth—in the knowledge of the way things are—it is essential to allow the authorship of the other to shine in experience and to play into our meanings.

Fundamental to authentic human consciousness is a loving allowance of things, themselves—loving in the sense of accepting and seeking to be true to their particularity and otherness. To allow something to show itself, as it itself is, is the essence of truth. The phenomenon of things appearing as they are is the ultimate reference for all discursive thought, understanding, and knowledge. It is what these latter are built upon and what they need to remain true to.[2] And the focus on full receptivity to nature foregrounds ways in which it is multisensory and bodied. The fluid awareness inherent in the constant delicate adjustments of bodily movement, too, through their tacit acknowledgement of things, brings these things into presence, sustains them. This reveals the perceiving subject as primordially corporeal and conscious: in its corporeality and consciousness it is an irreducible category of being[3] in which feeling and cognition, while formally distinguishable, existentially are so intertwined as to be co-constitutive of each other.

In participating in the being of things in this receptive way, embodied consciousness participates in a pre-predicative reality that is laden with meaning. Experience is no longer exclusively orchestrated according to instrumental purposes and discursively constructed conceptual schemes. Here the play of elemental powers that energize the natural world can be directly felt: powers such as those of birth and death, growth and decay, lightening and darkening, sound and silence, motion and stillness—each with intimations of fitting and unfitting response (Bonnett, 2015). Held in their sway, we are connected to the cosmos with a cogency that reveals abstract formulations as pale substitutes. The primary structure of reality becomes as much aesthetic and normative as it is mechanical. The intuited wholeness of the world arises from a sense of the latent potentiality out of which things in their "presencing" are constantly arising and falling back, much as when on a summer's day clouds carried along on some silent air bloom and disperse, are caught by the sun, stand forth in brilliance, and slowly melt into azure.

AUTHENTIC EDUCATION

Authentic education needs both to be grounded in an idea of authentic human being, and to elevate those experiences, discourses, and areas of study that reveal and nurture its potentiality.

In the foregoing, I have attempted to outline a view of authentic human being that is disclosed when attention is given to its potential for participation in the living presence of nature as a primordial reality. The following points were foregrounded:

- Engagement with nature, in its otherness, has shaped our senses, and is implicit in the logic of concepts that are generative of our form of consciousness;
- Engagement with nature entails a fully bodied, multisensory participation in its otherness that involves feeling as much as cognition, and receptivity to intimations of fitting and unfitting response;
- Here, in the letting be of things themselves, in a pregnant sense, a mutual sustaining occurs that is a primordial expression of truth.

Such considerations suggest that understanding the character of our relationship with nature is central to understanding the character of human being, and that the presence of a mutual sustaining as the vital centre of our authentic being requires properly to be recognized. But in our late-modern period, such recognition increasingly is occluded. Where traces remain, they are frequently disparaged—for example, as "naïve" or "romantic." Implicit in the account given has been a sense that the significances that emanate from direct experience of nature are constantly being effaced. With this effacement, what lies at the heart of authentic human being is veiled and subverted.

To be more explicit: today, thematized and theorized abstractions are increasingly offered as replacements for such original experience. When it comes to articulating reality, scientific and quasi-scientific narratives are allowed to dominate. Everyday language is either colonized by the metaphors that hold sway in these narratives or is disclosed as inadequate to the objectivity and mathematical nature asserted of the so-called "really real." While the approach of sociocultural studies constantly undermines confidence in experiences of the native otherness of the presencing of nature, the attitude of classical experimental science reifies nature in carefully defined categories and theories. Here the existential vitality and particularity of natural phenomena in their occurring become invisible. Such science increasingly articulates its knowledge in a form esoteric to everyday thought so that at the everyday level there is an important sense in which increasingly we know not what we do, nor the mechanisms by which we do it.

And there are other expressions of this withdrawal from original experience—for example, those now powerfully distilled in the narcotic hold in many lives of electronic connectivity, digitally constructed virtual realities, and, with technologies currently being developed, the likely burgeoning of digitally augmented realities. These all distance us from direct acquaintanceship with nature. In varying degree, they share the characteristics of Baudrillardian hyper-realities that have eddied off into a space of their own that no longer has a proper external reference. Such hyper-realities have lost connection with any profound reality such as raw nature or a deep sense of the human condition and exist only as pure free-floating simulations—much as when gossip takes off from reality and replaces it.

Earlier in this chapter I expressed the need to elucidate a conception of nature that would help to reveal the motives that lie behind attempts to erode its significance. Preoccupation with such hyper-realities seems to intimate (and to fan) a deep and destructive disdain for reality as given. In her Prologue to *The Human Condition*, Hannah Arendt (1998, pp. 1–3) noted how the launch of the first Sputnik was greeted in terms of its being "a first step towards escape from man's imprisonment to the earth." She commented that in late-modern times man (*sic*) seems "possessed by a rebellion against human existence as it has been given, a free gift from nowhere (secularly speaking) which he wishes to exchange, as it were, for something he has made himself."

It seems to me that what is being disclosed here is something that has long been on its way, and that with the ascent of modern technology threatens to occlude all else: the metaphysics of mastery. By this term I refer to a now overweening cultural motive to bring everything into the service of our self-given purposes. This motive installs us into a reality where essentially everything appears as an actual or potential resource—its very being consisting in its availability for manipulation, exploitation, and consumption. There is no room here for celebration of the fluid spontaneity and mystery of original nature, which now will show itself only as a passing curio or as an obstacle to be circumvented or overcome and brought to order. Hence, the rise of scientism that seeks to generalize the methods and language of science into all aspects of life, making all else appear as inconsequential, ultimately insubstantial: a frothy sideshow to the real business of life. Education provides a good example of this. Knowing, thinking, the curriculum, teacher–pupil relationships, school culture: under the metaphysics of mastery all need to be brought into line—modularized, micromanaged, outputs pre-specified.

In contrast, authentic education is implacably opposed to such scientistic pre-specification and the allied forms of instrumentalism that prevail in much current educational thinking and practice, and that are the bane of a free receptive engagement with the environment. Authentic education will give primacy to the ontological over the epistemological and will seek to develop/retrieve modalities of perception and knowing that better reflect the character of being in nature. This is suggestive of enhancing two intimately interrelated kinds of holism in education:

- The holism of the individual
- The holism of emplacement

The former refers to the individual conceived as embodied consciousness whose cognitive, emotional, volitional, aesthetic, moral, and spiritual dimensions are in organic interplay. The latter takes its start from the ecstatic nature of consciousness and authentic human being as environmental. It refers to the potentiality of an intimate relationship between self and a transcendent natural world of intrinsic agency and significances. This relationship involves a participation in the living presence of natural phenomena: the elemental otherness, mystery, and spontaneity that arise in the fluid emergence of things. Here truth and knowledge, with which education centrally should be concerned, are conceived less exclusively as a product of human ingenuity and calculation and more as an open receptivity to what announces itself. The implications of such a stance for the ethos and culture of educational institutions could hardly be more significant.

Notes

1. This is not to say that we should never seek to modify nature. It is simple to introduce the point that there is an important sense in which nature is normative and that recognition of this is relevant to responsible thought and action.
2. There are resonances here with some interpretations of Aristotle's *nous pathetikos* (*De Anima iii 4*)]. See, for example, Skúlason (2015).
3. Working in a very different (analytic) philosophical tradition, Peter Strawson (1964) comes to a similar conclusion with his notion of "persons."

References

Arendt, H. (1998). *The human condition* (2nd ed.). Chicago: University of Chicago Press.

Aristotle. (1987). *De anima*. London: Penguin Books.

Bonnett, M. (1999). Education for sustainable development: A coherent philosophy for environmental education? *Cambridge Journal of Education, 29*(3), 313–324.

Bonnett, M. (2004). *Retrieving nature: Education for a post-humanist age.* Oxford: Blackwell.

Bonnett, M. (2009). Education, sustainability, and the metaphysics of nature. In M. McKenzie, P. Hart, H. Bai & B. Jickling (Eds.), *Fields of green: Restorying culture, environment, and education* (pp. 177–186). Cresskill, NJ: Hampton Press.

Bonnett, M. (2012). Environmental concern, moral education, and our place in nature. *Journal of Moral Education Special Issue: Moral Education and Environmental Concern, 41*(3), 285–300.

Bonnett, M. (2015). The powers that be: Environmental education and the transcendent. *Policy Futures in Education, 13*(1), 42–56.

Brentano, F. (1995). *Psychology from an empirical standpoint.* London: Routledge.

Heidegger, M. (1972). *Being and time.* Oxford: Blackwell.

Heise, U. (2008). *Sense of place and sense of planet.* New York: Oxford University Press.

Husserl, E. (2001). *Logical investigations: Volume II.* London: Routledge.

Jickling, B. (1992). Why I don't want my children to be educated for sustainable development. *Journal of Environmental Education, 23*(4), 5–8.

Russell, B. (1959). *The problems of philosophy.* London: Oxford University Press.

Skúlason, P. (2015). The wildness of nature: Its significance for our understanding of the world. In P. Kemp & S. Froland (Eds.), *Nature in education* (pp. 81–89). Zurich: LIT Verlag.

Strawson, P. (1964). *Individuals.* London: Methuen.

Michael Bonnett has held senior teaching and research positions in the UK universities of Cambridge, London, and Bath. Formerly he was a Visiting Professor at the University of the Aegean. He has published widely in the field of philosophy of education and for some 20 years has taken a particular interest in the educational issues that arise from our understandings of nature and their implications for thinking about environmental concerns and sustainability education. His book *Retrieving Nature. Education for a Post-Humanist Age* was published in 2004 by Blackwell, and his edited collection *Moral Education and Environmental Concern* was published in 2014 by Routledge.

Environmental Education
After Sustainability

Lesley Le Grange

Abstract The idea, "after sustainability" has more than one meaning. It could mean "in pursuit of" (in imitation of) or "following in time." In this chapter I use both senses of "after sustainability." I firstly explore ways of rethinking sustainability by proposing the idea as a rhizome, as an empty signifier, and as the potentia of sustainability culture (a grass roots societal movement). This exploration is in pursuit of sustainability but offers radical alternatives to dominant discourses on sustainability. Second, I register the possibility of moving beyond the idea of sustainability, informed by an ontology of immanence, whereby both the subject becomes imperceptible and so too the idea of sustainability.

Keywords Sustainability · Sustainability education · Becoming imperceptible · Ontology of immanence

The word "sustainability" first emerged in German forestry management practices of the eighteenth century. It appeared in the *Oxford English Dictionary* for the first time in the year 1972. As an adjective, sustainability

L.L. Grange (✉)
Stellenbosch University, Stellenbosch, South Africa
e-mail: llg@sun.ac.za

© The Author(s) 2017   93
B. Jickling, S. Sterling (eds.), *Post-Sustainability and Environmental Education*, Palgrave Studies in Education and the Environment,
DOI 10.1007/978-3-319-51322-5_7

has been combined with many things such as agricultural practices, development, ecosystems, communities, societies, living, and even the entire Planet. Its combination with "development" is used most often and captured in the popular definition of the WCED (1987): "development which meets the needs of the present without compromising the ability of future generations to meet their own needs." This definition was an outcome of the Brundtland Commission that attempted to address the dissatisfaction of developing countries with the strong focus on conservation at the United Nations Conference on the Human Environment held in Stockholm in 1972.

Sustainable development has been a key focus of all major international conventions on environment since the Brundtland report and is encapsulated in the Millennium Development Goals (MDGs) of the United Nations as well as in policies of governments across the world. Sustainability has also been combined with education, and evidence of this is the Decade of Education for Sustainable Development (DESD 2005–2014) that was adopted by the UN General Assembly in 2002. However, sustainability/ sustainable development is a contested term, and so too its association with education. Moreover, the interpellation of sustainability into mainstream discourses has done little to arrest deepening poverty and the erosion of the world's biophysical base. The problematic nature of the term sustainability/sustainable development, including its association with education is the impetus for this book on post-sustainability. In this chapter, I shall refer to "after sustainability" instead of "post-sustainability" because the former enables me to both rethink sustainability (education) and to explore possibilities of transcending the term.

The idea, "after sustainability," has more than one meaning. It could mean "in pursuit of sustainability" (in imitation of) or "following sustainability in time." I use both senses of "after sustainability" in this chapter. I will invoke the former meaning to rethink sustainability without jettisoning the term and the latter meaning to explore ways of transcending the term. Before exploring the two ideas I shall first discuss the contested nature of sustainability as well as difficulties concerning its relationship with education. Second, I shall explore ways of rethinking sustainability: by thinking of the idea as a rhizome, as an empty signifier, and as the *potentia* of sustainability culture (a grass roots societal movement). This exploration is "in pursuit of sustainability" (the first sense of "after sustainability"), but offers radical alternatives to dominant discourses on sustainability and draws on the ideas of scholars such as Deleuze, Laclau and Parr. Third, I shall register the possibility of moving beyond the idea of sustainability that is informed by an ontology of immanence, whereby both the subject and the concept sustainability become imperceptible. By an ontology of immanence, I mean that the world (reality) does not comprise separate self-contained substances, but that everything that exists ("living" and "non-living") is a modification of life. The rock and the human are actualized forms of the same life force, a force that does not exist outside of the rock or the human. Action in the world is driven by life itself—by that which is within, and that connects everything that exists in the cosmos—not by a substance or idea outside of life. The upshot of an ontology of immanence is the disappearance (the becoming imperceptible) of the subject—the disappearance of the atomized subject that transcends the world and comes to know it by distancing itself from the world. And, it also concerns the dissolution of any notion that "transcends" life itself. In other words, the subject becomes ecological rather than a transcendent human, and environmental action becomes simply doing without invoking a transcendent construct such as sustainability. By this I mean that ethical action is driven by life itself (the creative power of life) and not by an external force such as an idea, a policy, a goal, or a principle.

Sustainability (Education): A Contested Terrain

Sustainability/sustainable development has been the subject much contestation. It has been both eulogized and demonized. But, there are difficulties with the term and some of the criticisms levelled against sustainability are that: it has internal contradictions, it manifests epistemological difficulties, it reinforces a problematic anthropocentric stance, it has great appeal as a political slogan, it is a euphemism for unbridled economic growth, it is too fuzzy a term to convey anything useful, and it does not take into consideration the asymmetrical relation between present and future generations (for detail see Bonnett, 1999, 2002; Le Grange, 2008; Stables & Scott, 2002). Irwin (2008) also notes that sustainability has been taken up in neoliberal discourses and permeates multinational corporations, pan-global organizations, national governments, education policy, institutions, and curriculum. In a similar vein, Parr (2009) suggests that sustainability has been hijacked by the military, government, and the corporate world. Furthermore, Le Grange (2013) argues that the notion of "needs" reflected in the popular definition of sustainable development should be understood in the context of the emergence of "needs" as a political discourse in late capitalist society—that "needs" is a political instrument.

The relationship between education and sustainability is also a contested terrain. Environmental educationists have widely differing views on the relationship between sustainability and environmental education. As Sauvé (1999) points out, for some, sustainable development is the ultimate goal of environmental education, thus the term *environmental education "for" sustainable development* (EEFSD). For others, sustainable development encompasses specific objectives that should be added to those of environmental education, thus the expression *education for environment "and" sustainable development* (EFE and SD). For others still, environmental education inherently includes *education for sustainable development*, thus the use of both terms is tautological. Scholars such as John Huckle and John Fien argue that it is imperative to educate for sustainable development given unprecedented levels of environmental degradation and social injustices. Their perspective is informed by a socially critical view of environmental education. The idea of educating for sustainable development has also been the approach of the Decade of Education for Sustainable Development (DESD). However, in a recent analysis of four DESD products, Huckle and Wals (2015, p. 502) conclude that the Decade has been "business as usual" as far as challenging neoliberalism and in promoting global education for sustainability and citizenship (GESC).

Other scholars such as Jickling (1997) have troubled the idea of "educating for" sustainability/sustainable development arguing that such an approach suggests an instrumentalist view of education. In fact, Jickling goes as far as to say that education for sustainability is anti-educational and tantamount to indoctrination—that "education for" anything, implies that it must strive to be "for" something external to education itself (p. 95). I aver that "sustainability education" is a more useful signifier for a relationship between sustainability and education because it is non-instrumentalist. Sustainability education does not signify an a priori image of sustainability nor defines what the education pathway towards achieving sustainability should be. Instead it opens up possibilities for critical discussions on sustainability and suggests a process that is always in-becoming.

Huckle and Wals' finding, following their analysis of four Decade of Education for Sustainable Development products, is unsurprising. All processes that follow an instrumentalist logic manifest a commitment to transcendence which Nietzsche argued is one of the great errors in Western thought. Transcendence is the idea that there are two ontological substances and that the one transcends the other. An ontology of transcendence is based the idea that reality is comprised of separate self-contained substances that interact to bring about change/difference in the world. Moreover, that there is a God, a human subject, a soul, a form, or idea that transcends life itself. It is a notion that underpins Plato's Forms, Decartes' dualism, Newton's physics, Hegel's dialectic, Marx's superstructure that creates ideological relations, and so on (Wallin, 2010). Deleuze and Guattari (1994) assert that transcendence is the belief in the existence of a substance/thing beyond empirical space, power, or existence (ontological being). It is the commitment to transcendence that has separated humans from nature (causing nature's destruction) in the sense that the human subject transcends nature, is able to distance itself from it, know it objectively, and control and manipulate it for its own ends. The commitment to transcendence has also informed an education system that has reinforced dualisms, such as theory/policy and practice, teaching and learning, and so on. Transcendent thinking is not only evident in conservative positivist approaches to education but also in critical pedagogy informed by Marxist thinking. Bowers (1980, p. 282) points out that proponents of critical pedagogy frame socialism and capitalism in a dualistic logic of right/wrong, truth/illusion, and salvation/damnation.[1] A commitment to transcendence is evident in the approach to education for sustainability/sustainable development (ESD), because it is premised on the idea that some notion of sustainability transcends the immanent process of education—such transcendent notions of sustainability are usually captured in the form of goals, aims, and objectives. Against this background, I now explore ways of rethinking sustainability (education) that moves in the direction of immanence (anti-transcendence).

Rethinking Sustainability (Education)

This section explores ways of rethinking sustainability so as to generate alternative possibilities to its framing within dominant discourses. The exploration is "in pursuit of sustainability" in the sense that sustainability (education) is rethought without jettisoning the term.

First, in my own work I have argued that sustainability could be a carrier of alternative possibilities if viewed rhizomatically instead of arborescently (Le Grange, 2011, p. 744). The latter refers to conceptions of knowledge as hierarchically ordered branches of a central stem/trunk rooted in firm foundations. The former refers to chaotically complex networkings of stems interconnecting the shoots of some grasses (Sellers, 2006). An arborescent view of sustainability (education) holds that it is centred in global neoliberal discourses that branches in tree-like fashion to the periphery (local). A rhizomatic view of sustainability (education) decentres it, producing a distributed knowledge system that opens up pathways for marginalized knowledges including Indigenous ones. Understood in this way, sustainability education connects the ideas, tools, and skills of all participants involved (community members, academics/teachers, and students) in multiple ways to produce "new" knowledge in "new" knowledge spaces. Deleuze and Guattari (1987) remind us that the rhizome has no points or positions, such as those found in a structure, tree, or root—there are only lines. Lines enable proliferation in all directions to form assemblages. Sustainability (education) therefore could be understood as an assemblage, meaning that it increases its dimensions of multiplicity, and necessarily changes its nature as it expands its connections.[2] Viewed in this way, sustainability is rescued from the normalizing, homogenizing, and domesticating effects produced by an arborescent view of the term. Put simply, sustainability (education) as a rhizome connects in multiple ways with people and the more-than-human world and learning involves understanding the interconnectedness of humans and humans, and humans and the more-than-human-world, and about how new connections might be generated.

Second, sustainability could be viewed as an "empty signifier." This view is posited by Brown (2015) who draws on the work of Ernesto Laclau. For Laclau (1996) an empty signifier is a "signifier without a signified." It is not a word without meaning but concerns the possibility of signifying the limits of signification itself. Brown (2015, p. 3) writes:

This "limit" refers not to a neutral, empirical boundary, as such a boundary could itself be signified and thus be incorporated into the signifying system. The limit in question is rather what has been excluded from the discourse. It is a "radical" limit. . . . Since what is outside such a limit cannot be signified except through inclusion into the signification, the only way in which this limit can "appear" is through the interruption or failure of the very process of signification.

Brown (2015, p. 10) argues that empty signifiers stand in the gap when there are mutually incompatible discourses—discourses which are antagonistic. Discourses are antagonistic when they cannot be incorporated within a particular system of signification. Antagonistic relations result when there is something that the discursive system is unable to hold and leads to the dislocation of the identity of those who constitute the relation. Put simply, an empty signifier holds what a particular discourse excludes, what a signification system cannot incorporate. In the context of our discussion, when dislocation occurs it brings into sharp focus the untenable futures a discourse is producing. For example, the untenable future that might have been produced by the strong focus on conservation at the UN Conference on the Human Environment held in Stockholm in 1972 was brought into sharp focus when this discourse was challenged in the 1980s by members of the developing world who argued for development to continue in their countries so as to overcome poverty and related concerns. In this instance, as empty signifier, sustainability functioned as a tool to hold these antagonistic discourses because each of these discourses was unable to incorporate the other. Likewise, undesirable futures that would result from the continued use of fossils fuels (forming part of the economic rationalist discourse) have been brought into sharp focus by the climate change discourse—a discourse antagonistic to the economic rationalist one.

Some scholars such as Swyngedouw (2010) argue that sustainability is inherently reactionary therefore it has become co-opted and hegemonized into narrow discourses such as sustainable development. In response, Brown (2015) avers that the co-optation and hegemonization of sustainability into a narrow discourse of sustainable development could be challenged by discourses antagonistic to it—discourses outside this dominant discourse. As Brown (2015, p. 13) writes:

> Given the fact that [the sustainable development discourse] operates cynically and does not address the fundamental issues that created the conditions for sustainability politics in the first place, there will always be a considerable "excess" within the social field that sustainability cannot neutralise. As an empty signifier, "sustainability" represents that impossibility and the aspiration to attain it.

The dominant discourse of sustainable development is untenable because it relies on significant exclusions—exclusions captured by a rhizome of disparate social groups/movements: feminists, upstanding citizens, vegans, anarchists, communists, right-wing groups, environmentalists, to name a view. As an empty signifier, sustainability stands in the gap for what narrow discourses such as sustainable development exclude. As Brown (2015, p. 17) writes: "as an empty signifier, sustainability allows these multiple ruptural points to be condensed in a generalised concern for the future." In short, as an empty signifier, sustainability has the potential of functioning as a tool for radical politics, because it brings into focus what is excluded from dominant discourses.

Third, in her book *Hijacking Sustainability*, Adrian Parr (2009) suggests that there is an alternative conception of sustainability to its co-opted form by governments, the military, and the corporate world. She notes the need to distinguish between culture that functions as a point of disequilibrium and insurgency, and the mediated form of culture that functions as a point of control and order. The former she suggests is used to enhance life and the latter to limit life. The unmediated (or less mediated) culture Parr calls, "sustainability culture." The power of sustainability culture is *potentia* (the power of the multitude) and its presence curbs the power of the state and the corporate world, that is, *postestas* (the power of the sovereign). Sustainability culture taps into the creative and productive energies of *potentia*, inviting us "to imagine and design alternatives to how a culture is produced, disseminated, and consumed" (Parr, 2009, p. 165). Moreover, it is optimistic insofar as it encourages us to work for a future that is based on the interest of the common good rather than on maximizing profits. Sustainability culture aspires to create processes that affirm the vitality and dynamic materialism of life as these imbue life—this vital materiality is the ontological energy shared by all of life (p. 165). It (re)generates life by tapping into what is immanent to *potentia*, producing what is unimaginable but within the limits set by life itself. As Parr (2009, p. 165) writes: "Yes life has limits—Earth's metabolism can gulp down only so much of our waste, and Earth can recycle only a finite amount of the toxins industry spews into the atmosphere." Sustainability education that is intimate to sustainability culture taps into the creative energy of *potentia* and connects (people to people, people to nature, discipline to discipline), generates newness, and encourages transversal thinking. It counteracts manifestations of *postestas* in education such as disciplinary thinking/learning, predetermined outcomes or objectives, standardized tests, and so on.

These three approaches to rethinking sustainability are not mutually exclusive and offer alternatives to its co-opted and hegemonized forms such as sustainable development. However, to lesser or greater degrees, all three approaches descend into a form of what Meillassoux (2008, p. 5) terms "correlationism." Correlationism is a term used by a recent development in philosophy called speculative realism, which as the name suggests, concerns a return to speculating the nature of reality independently of human thought. Put simply, correlationalism means that reality appears only as the correlate of human thought—the limit of correlationalism is why conventional continental philosophy might be considered to be anthropocentric. Alongside speculative realism, new materialisms have developed as an interdisciplinary field of inquiry produced by feminist scholars. In short, these scholars share the view that humans are not only socially, discursively, and linguistically constructed but also materially constructed. By "material," it is meant that human beings are made of the same physical materials as the non-human (more-than-human) world and that all human systems (including systems of thought) are underpinned by material flows. Both speculative realists and new materialists are largely opposed to naïve realism but hold that continental philosophy (phenomenology, structuralism, post-structuralism, deconstruction, and postmodernism) has limits in responding to the current ecological crisis and the advancement of technology that is blurring the boundaries between human and machine. It is the (re)turn to realisms and the ontology of immanence alluded to earlier that serve as the basis for my discussion on the "becoming imperceptible" of sustainability (education).

The three approaches to rethinking sustainability discussed are illustrative of environmental education "after sustainability" that is, in pursuit of or in imitation of sustainability. In these cases sustainability education is rethought but without discarding the idea of sustainability. Next, the other sense of "after sustainability" is discussed, that moves beyond the idea of sustainability and overcomes the problem of correlationalism.

THE BECOMING-IMPERCEPTIBLE OF SUSTAINABILITY (EDUCATION)

An ontology of immanence relates to the oneness of all entities in the cosmos, "living" and "non-living"—that all entities are actualized forms of the flows and intensities of life's creative power—the human and the rock are both modifications of life. Deleuze (1988) presents the ontology of immanence and the creative power of life as an ethical principle in his book on Spinoza, an ethics called an immanent ethics. Žakauskaitė (2015) argues that the creative power of life functions as an ethical principle in two senses: first, it opposes any transcendent values and follows immanent rules implicit in the modes of existence; second, the creative power of life dissolves the model of subjectivity and at the same time the powers of subjectification. In this way, the conditions of "the ethics of becoming-imperceptible" are created. Becoming-imperceptible means the disappearance of the atomized subject—rather than subjectivity being individual, it is ecological. It signifies a shift from the arrogant "I" (of Western individualism) to the humble "I"—to the "I" that is embedded, embodied, extended, and enacted. As Braidotti (2013, p. 137) writes:

> Becoming-imperceptible marks the point of evacuation or evanescence of the bounded selves and their merger into the milieu, the middle grounds, the radical immanence of the earth itself and its cosmic resonance.

The disappearance of the individual self that characterizes becoming-imperceptible overcomes the problem of correlationalism that I described earlier. In becoming-imperceptible the cosmos or the earth is not reduced to human thought, but human thought is instead bent by the earth/cosmos. The "human" does not simply inhabit the world/earth but is inhabited by it. In the education process the unit of reference is not the individual subject but an assemblage produced by processes immanent to the creative power of life. The earth is not a stage on which pedagogy is performed but the performance of pedagogy is bent by the earth.

As mentioned, an immanent ethics is opposed to any transcendent values, or universalizable objective truths and principles[3] and therefore is in contrast to moral theory. As Smith (2012, p. 221) writes:

> Life does not function . . . as a transcendent principle of judgement but as an immanent process of production or creation; it is neither an origin nor a goal, neither an *arche* or a *telos*, but a pure process that always operates in the middle, au milieu, and proceeds by means of experimentations and unforeseen becomings.

An immanent ethics does not function with a transcendent value or goal such as sustainability in mind. As the three interlocking dimensions of environment, the biophysical, social, and mental (Guattari, 2001) are being destroyed, the question is not, "What must I do?" (or how must I live or learn) but "What can I do?" or as Braidotti (2013, p. 259) suggests, "Just do it."

When an immanent ethics functions, sustainability itself becomes imperceptible ensuring that sustainability cannot become an order-word (*mots d'ordre*) or a political slogan. Braidotti (2006, p. 260) elaborates, "There is no assurance . . . of a teleologically ordained trajectory, just the urge to get on with it, just do it, though the final destination may not be very clear. All that matters is the going, the movement." For Deleuze (1988, p. 23), "good" are those modes of existence that expand our powers to act (to just do) and "bad" are those that keep us in a state of passive slavery.

An education informed by an immanent ethics opens up pathways for students to increase their powers of acting, to express their generosity, and to love the world (all of life)—and it is an invitation to dance (to just do) (Braidotti, 2006, p. 259). Humans' abilities or capacities to dance or swim come from the power that is within us and not from some force outside of us. Nobody, for example, teaches a dog to swim, the dog just swims. So too, are our capacities to live, to love the world, and to connect positively to everything in the cosmos, cannot be taught. It is when the power within us is suppressed by that which is external to life's creative power that we see the erosion of Guattari's three ecologies, *mental, social,* and *environment.* Smith (2012, p. 285) writes about an immanent ethics in the following way:

> The fundamental question of ethics is not "What *must* I do?" (the question of morality) but rather "What *can* I do?" Given my degree of power, what are my capabilities and capacities? How can I come into active possession of my power? How can I go to the limit of what I "can do."

The role of the educator is therefore not to tell the student what they *must* do; to follow what is prescribed in terms of content, aims, and objectives, but to help students to unlock the creative power within them so that they can ask, "What can I do?" "What can I learn" and/or "How can I learn?" An invitation to dance is an invitation to release our inner capabilities and capacities produced by the creative power of life itself so as to collectively make the world a different (better) place. The idea of unlocking the power of life might be viewed by some as being romantic but in view of contemporary theory's obsession with "political violence, wounds, pain, and suffering" (Braidotti, 2010, p. 142) and peoples' fears of the effects of potential environmental disasters, terrorism, warfare, genetic manipulation, and so forth, there is a need to experiment (theoretically too) with ways of affirming life, without negating the importance of mourning and having empathy with suffering.

In South Africa, with its challenges of drought, poverty, epidemics, and pandemics, an ontology of immanence can be harnessed by invoking words derived from aphorisms found in African languages such as the Shona word *Ukama* (relatedness of everything in the cosmos) and the *isiXhosa* word, *Ubuntu* (because we are, I am) (see Le Grange, 2012a, 2012b). Both these words express the oneness of everything in the cosmos and the imperative to care for the "living" and "non-living" (the more-than-human world). This imperative to care is within us as part of our being in the world, but becomes eroded or obfuscated by our cultures, schooling, and limitations of society.

SOME PARTING THOUGHTS

In this essay I used two senses of the idea "after sustainability (education)." The first sense was in imitation of sustainability by exploring ways of rethinking sustainability without discarding the term. Radical alternatives to dominant discourses on sustainability were discussed, which drew on insights from scholars such as Deleuze, Laclau, and Parr. The second sense of "after sustainability" opens up ways of moving beyond the idea of sustainability, informed by an ontology of immanence, whereby both the subject becomes imperceptible and so too the idea of sustainability. The subject becomes ecological rather than a transcendent human and environmental action becomes simply doing without invoking a transcendent construct such as sustainability. A lot more can be said on the notion of post-sustainability. My aim was to add to a conversation that might open up pathways that could take us beyond sustainability so that we can "just do," that is, to unlock our collective capabilities and capacities to make a difference (or rather be a difference); not by imposing anything on the world but in releasing the creative power which is (with)in all of life (rock, tree, river, human, etc.).

NOTES

1. Dualism is linked to transcendence in the sense that dualism is based on the premise that an entity/substance exists only relation to another entity/ substance external to itself.
2. At Stellenbosch University, South Africa we have, for example, seen the deterritorialization of some disciplines located in disparate faculties and the

emergence of a transdisciplinary network of scholars and the development of a transdisciplinary doctoral programme involving academics from all faculties within the university. Moreover, we have seen lines of flight from this network to form connections with other research organizations and local communities. An example of the production of "new" knowledge in a "new" knowledge space is the Enkannini (local community of shack dwellers) iShack project which is an assemblage of community members, university academics, students, and community-based organizations working together among other things: improve insulation of shack buildings using recycled material; introducing biogas digesters utilizing human solid waste to enable cooking from biomethane; using grey water flushing; and producing an off-grid solar home system.

3. Wallin (2010, p. 25) importantly points out that the premise of one ontological substance (not two) does not mean that we are left with a uniform plane that denies difference. He writes, "The attributes of substance accessible to human consciousness include thought and extension, both of which Deleuze conceptualizes as an unlimited finity... [T]he attribute is not attributed by a transcendent substance, but is rather one of an infinite number of ways a substance might be expressed." Students' powers of acting can of course be expressed in opposite ways, where they act selfishly or show hatred to the world. This happens when the creative power within is colonized through arrogance, institutional regimes, etc.

REFERENCES

Bonnett, M. (1999). Education for sustainable development: A coherent philosophy for environmental education? *Cambridge Journal of Education, 29*(3), 313–324.

Bonnett, M. (2002). Sustainability as a frame of mind-and how to develop it. *The Trumpeter, 18*(1), 1–9.

Bowers, C. (1980). Curriculum as cultural reproduction: An examination of the metaphor as carrier of ideology. *Teachers College Record, 82*(2), 267–290.

Braidotti, R. (2006). *Transpositions*. Malden, MA: Polity Press.

Braidotti, R. (2010). Powers of affirmation: Response to Lisa Baraitser, Patrick Hanafin and Clare Hemmings. *Subjectivity, 3*(2), 140–148.

Braidotti, R. (2013). *The posthuman*. Malden, MA: Polity Press.

Brown, T. (2015). Sustainability as empty signifier: Its rise, fall and radical politics. *Antipode*, 1–19. doi:10.1111/anti.12164

Deleuze, G. (1988). *Expressionism in philosophy: Spinoza* (Tom Conley, Trans.). New York: Zone.

Deleuze, G., & Guattari, F. (1987). *A thousand plateaus: Capitalism and schizophrenia* (Brian Massumi, Trans.). Minneapolis: University of Minnesota Press.

Deleuze, G., & Guattari, F. (1994). *What is philosophy?* (Hugh Tomlinson and Graham Burchell, Trans.). New York: Columbia University.

Guattari, F. (2001). *The three ecologies* (Ian Pindar and Paul Sutton, Trans.). London: The Athlone Press.

Huckle, J., & Wals, A. E. J. (2015). The UN decade of education for sustainable development: Business as usual in the end. *Environmental Education Research, 21*(3), 491–505.

Irwin, R. (2008). "After Neoliberalism": Environmental education to education for sustainability. In E. Gonzalez-Gaudiano & M. A. Peters (Eds.), *Environmental education: Identity, politics and citizenship* (pp. 171–193). Rotterdam: Sense publishers.

Jickling, B. (1997). If environmental education is to make sense for teachers, we had better rethink how we define it. *Canadian Journal of Environmental Education, 2*, 86–103.

Laclau, E. (1996). *Emancipation(s)*. London: Verso.

Le Grange, L. (2008). Towards a language of probability for sustainability education (South) Africa. In E. Gonzalez-Gaudiano & M. A. Peters (Eds.), *Environmental education: Identity, politics and citizenship* (pp. 207–217). Rotterdam: Sense publishers.

Le Grange, L. (2011). Sustainability higher education: From arborescent to rhizomatic thinking. *Educational Philosophy and Theory, 43*(7), 741–754.

Le Grange, L. (2012a). Ubuntu, ukama, environment and moral education. *Journal of Moral Education, 41*(3), 329–340.

Le Grange, L. (2012b). Ubuntu, ukama and the healing of nature, self and society. *Educational Philosophy and Theory, 44*(S2), 56–67.

Le Grange, L. (2013). The politics of needs and sustainability education in South Africa. In R. Stevenson, M. Brody, J. Dillon, & A. E. J. Wals (Eds.), *International handbook of research on environmental education* (pp. 108–114). New York: Taylor & Francis.

Meillassoux, Q. (2008). *After finitude: An essay on the necessity of contingency* (Ray Brassier, Trans.). New York: Continuum.

Parr, A. (2009). *Hijacking sustainability*. Cambridge, MA: The MIT Press.

Sauvé, L. (1999). Environmental education between modernity and postmodernity: Searching for an integrating educational framework. *Canadian Journal of Environmental Education, 4*, 9–35.

Sellers, W. (2006). Review of technology, culture, and socioeconomics: A rhizoanalysis of educational discourses by Patricia O'Riley. *Transnational Curriculum Inquiry, 3*(1). Retrieved from http://nitinat.library.ubc.ca/ojs/index

Smith, D. W. (2012). *Essays on Deleuze*. Edinburgh: Edinburgh University Press.

Stables, A., & Scott, W. (2002). The quest for holism in education for sustainable development. *Environmental Education Research, 8*(1), 53–60.

Swyngedouw, E. (2010). Apocalypse forever? Post-political populism and the spectre of climate change. *Theory, Culture, and Society, 27*(2/3), 213–232.

Wallin, J. J. (2010). *A Deleuzian approach to curriculum: Essays on a pedagogical life*. New York: Palgrave Macmillan.

WCED (World Commission on Environment and Development). (1987). *Our common future*. Oxford: Oxford University Press.

Žakauskaitė, A. (2015). Deleuze and Beckett towards becoming-imperceptible. In S. E. Wilmer & A. Žakauskaitė (Eds.), *Deleuze and Beckett* (pp. 60–77). New York: Palgrave Macmillan.

Lesley Le Grange is Distinguished Professor in the Faculty of Education at Stellenbosch University, South Africa. He is a former Chair of the Department of Curriculum Studies and Vice Dean (Research) of the Faculty of Education at Stellenbosch University. Lesley teaches and researches in the fields of environmental education, research methodology, science education, and curriculum and assessment. He has 186 publications to his credit and serves on editorial boards of seven peer-reviewed journals. Leading international journals in which he has published in recent years include: *Educational Philosophy and Theory, International Review of Education, Journal of Curriculum Studies, Journal of Moral Education*. Lesley has contributed to several international handbooks on/in education, the most recent the Springer, *International Handbook of Interpretation in Educational Research Methods* (2015). He is a member of the Accreditation Committee of the Council on Higher Education in South Africa and Vice President of the International Association of the Advancement of Curriculum Studies (IAACS). In environmental education his current research interests are: critically "analysing" sustainability and its relationship to education; developing *Ubuntu* as an environmental ethic and exploring its implications for education. His contribution to this book draws on insights from two decades of research that he has conducted on sustainability education.

PART IV

Education Through Action

Education as Life

Lucie Sauvé

Abstract Through my immersion in citizen resistance movements against invasive "development" projects, and my exploration of diverse ecosocial creative initiatives, I have observed the ongoing processes of knowledge construction and very often, the emergence of a collective and powerful intelligence in the mobilized groups. Noting the gap between school and such contexts of rooted and meaningful learning, some questions emerged: how can formal education build on these informal leaning dynamics so as to empower youths/citizens willing and capable of contributing to social debates and transformations for better living together within our *oikos*? In this perspective, how can the barriers between formal, non-formal and informal learning contexts be broken? More specifically, beyond any globalizing educational projects, what is/could/should be the contribution of environmental education to our human journey in this world?

Keywords Environmental education · Ecocitizenship · Ecosocial transformation · Ecological democracy · Grounded learning

L. Sauvé (✉)
Université du Québec à Montréal (UQAM), Québec, Canada
e-mail: sauve.lucie@uqam.ca

© The Author(s) 2017
B. Jickling, S. Sterling (eds.), *Post-Sustainability and Environmental Education*, Palgrave Studies in Education and the Environment,
DOI 10.1007/978-3-319-51322-5_8

111

There have been terrorist attacks in Paris, Baghdad, Ouagadougou, Mogadishu, Brussels, Orlando, Nice, Berlin, Istanbul, and so many more. Painful tides of migrants try to reach the beaches of exile. Fiscal evasion has become a standard smart practice. Clowns climb onto the political stage. The "left" is dramatically weakened or pushed out in Latin America and elsewhere. Climate change is irredeemably going on, drying, burning, melting, flooding, devastating... while the "extreme" oil and shale gas industry tries to gain hold and the pipeline octopus finds and forces ways to spread its giant limbs. What next? The Arab Spring and other seasons of hope will not have bloomed?

Here, I could stop writing this text: *there is no point, it is too late!* But I will not. I received a good dose of hope through my deep immersive experience in social movements and my exploration of so many coura- geous, grounded ecosocial projects. If the global scale is—for here and now—out of reach, the sphere of local and regional action is largely open to concrete involvement, and there grows a whole rhizomatic world of meanings and skills patiently and courageously emerging, and transform- ing the landscape of our contemporary humanity.

> Apathetic and resigned, the civil society? In reality, it has never been so lucid, so intelligent. It is discovering its collective strength, expresses an articulated critical thinking that brings the end of the neoliberal culture hegemony and proclaims its demands for real change. (Manier, 2012, p. 302)

Despite the unreasonably heavy tasks of resistance and of recreation of the world, despite the difficulties and setbacks, protests, marches, and other collective strategies still get organized at every crisis. Committed citizens claim the legitimacy of their democratic and active participation to governance, to the decisions that concern them. Transformative projects are also blooming creatively in the fields of energy, food, housing, and other dimensions of our individual and community lives. A collective intelligence—and even more, a citizen intelligence—is being built, with courage. Despite the present global failures (including all forms of social and ecological violence), the upsurge of a major cultural change is taking shape and getting stronger, as much from ethical and political perspectives as from strategic and economic ones, aiming to recover and celebrate the dignity of life.

This situation challenges the diverse sectors of education, but parti- cularly school education. The current politics of formal education (from curriculum design to teacher education and classroom settings) maintain a huge gap between schooling and different social contexts where rooted, strategic, and meaningful learning can occur. Too often, school stands as an island—offshore—where youth remain waiting for "real life," learning things that "will be useful later." In order to really contribute to an "educational society" (Delors, 1996), and better still, to an emancipatory "learning society," how could school derive inspiration from the informal leaning dynamics that are taking shape in the movements of resistance and ecosocial initiatives from whence societal transformation is slowly emerging? How could school support, value, complete, and transfer these learning processes? How can we foster gateways between school and the other different learning actors and contexts (community organizations, NGOs, museums, parks, media, etc.) so as to stimulate and sustain citizen competencies and ecosocial involvement? These are huge questions that certainly cannot be discussed in depth in this text. But it is important to acknowledge their relevance and contribute to reflection on this theme. We will examine here the contribution of the rich theoretical and pedagogical heritage of environmental education, as an ontological and political process towards ecocitizenship.

What Do We Learn Through Social Action?

I like to tell this story—similar to so many other situations where citizen commitment becomes the crucible of essential learning. In 2013, people from the Saint Lawrence estuary heard about the TransCanada Energy East pipeline project that planned to build a giant oil harbour at Cacouna, a small town along the river. Reacting to a strong feeling of territorial usurpation from a foreign private company, a citizen mobilization quickly emerged in the affected area, as well as through the whole watershed. Not only did the project jeopardize this Québec fluvial artery, but also the proposed harbour was situated right in the reproduction zone of the Beluga, an endangered marine mammal. On the protesters' signs, one could read sentences as "Don't touch my river!" "The river flows in our veins," "I am Beluga!" A rooted and emerging collective consciousness of territorial identity thus appeared, as part of what Mitchell Thomashow (1996) calls an ecological identity, a sense of belonging to a life system.

Through such dynamics as resistance, citizens learned—once again—that in the context of the current "governance," centred on political and economic alliances and backed by a complacent or incoherent legislative system, it is up to civil society to assume the difficult and very demanding role of becoming a critical vigil, of fighting *against* or *for* projects, programmes, or policies that affect their lives, places, and territories. The ecological argument, the one for the preservation of the integrity of the ecosystems, as well as the argument of the common goods (health, security, auto-determination, etc.) seems to have no importance for policymakers, except when they are upheld by popular discontent, requests, and claims and they become embarrassing for the elected politicians. So with regard to issues like the Cacouna harbour project, in an emergency context, mobilized citizens have to very quickly get informed, understand, build up an argument, communicate, debate, plan strategically, protest, act, and propose. They have to learn how to work together, share knowledge, experience, expertise, and resources. Collaboration is challenging, but this is how citizen intelligence can be developed, as a global comprehension and a collective capacity to act over a very political situation. And these elements of learning and empowerment turn out to be transferable to other socio-ecological struggles.

The social mobilization against the oil project not only has allowed the Beluga habitat to be to preserved, but also—like so many other resistance movements against invasive projects of this kind—it has contributed to the creation of social links and the formation of a vigilant ecocitizenship, now more aware of the necessity of "free, prerequisite, and enlightened consent" of populations facing "development" initiatives in their territory. Through this courageous debate, the desire for a renewed democracy based on transparency, deliberation, collaboration, cooperation, participation, and direct action was expressed.

From such experiences, we have understood that denouncing and resisting is not enough; one has also to create. Resisting is creating (Aubenas & Benasayag, 2002), thus opening up huge and meaningful working for ecosocial initiatives, innovations, and practices. This is why we need to hear more and more about the inspiring stories of community gardens, workshops to share tools and skills, networks of solidarity between farmers and "eaters," eco-villages, and so many others, where resistance to capitalism "becomes an act of creation, able to produce through itself other values" (p. 114).

FORMAL EDUCATION AND ECOSOCIAL TRANSFORMATION

But how can formal education get inspired from the extraordinary learning processes that mobilized citizens' experience together through the dynamics of resistance and collective projects? How can schools invite, prepare, and accompany youth to take part in the current ecosocial transformation movement? How can education value young peoples' creative force and respond to their quest for meaning and desire for action?

The idea is not to shape "future citizens," but to consider and accompany youth as fully fledged players in our world. In her book *Children, Citizenship and Environment*, Bronwyn Hayward (2012) insists on the role of children as social actors in their community. It is not a matter of teaching political science, she writes, but to offer children the possibility of taking consciousness of their place and role in the collective life, and to experiment with active democratic processes. Children need to recognize that "ecological" problems are closely linked to problems of violence, poverty, injustice, and inequity. Our role as educators is to invite them to talk about their daily living, to clarify their reality, and to experiment and understand how ordinary people (like their parents, their teachers, their neighbours, and themselves . . .) can "act together in free collaboration to achieve extraordinary change" (p. 2). Since the environment is a "common good," one must learn to live with commitment and democracy in order to contribute to the transformations of socio-ecological realities, starting with those that concern us the most, right here, in our living places. "We need to support young citizens as they discover the art, craft and passion of active ecological citizenship" (p. 16).

This critical requirement for democracy, cooperative learning, and collective construction of meaning and action projects echoes John Dewey's pedagogy. It also echoes Noam Chomsky's hope that human beings, because of their "instinct of liberty," are able, if their development is not completely impeded, to raise up victoriously against whatever oppresses them (Baillargeon, 2010, p. 10).

Edgar Morin (2014) also wishes to transform education towards autonomy and liberty of choice through different opinions, theories, and philosophies. He insists on the importance of learning "how to know," recognizing that knowledge is always translation and reconstruction. This metacognitive posture helps us recognize the risk of mistakes and illusions, of partial or biased knowledge. As he says, "To live implies the need, in order to act, to access relevant knowledge that is not mutilated, nor mutilating, that replaces all objects or events in their context and that apprehends them in their complexity," while recognizing zones of uncertainty (p. 20).

And beyond these epistemological aspects, education for social transformation finds its major roots in Paulo Freire's politico-pedagogical proposal for freedom, hope, and love:

> Love is the basis of an education that seeks justice, equality, and genius. If critical pedagogy is not injected with a healthy dose of what Freire called "radical love," then it will operate only as a shadow of what it could be. [...] Critical pedagogy uses it to increase our capacity to love, to bring the power of love to our everyday lives and social institutions, and to rethink reason in a humane and interconnected manner. [...] A critical knowledge seeks to connect with the corporeal and emotional in a way that understands at multiple levels and seeks to assuage human suffering. (Kincheloe, 2005, p. 3)

These educational preoccupations of epistemological, ethical, affective, strategic, and political order expressed by the different authors mentioned above—as examples of this type of discourse—converge towards the necessity of engaging a major shift of our education systems. Indeed, the integration of cross-curricular or transversal dimensions that have characterized curricular reforms throughout the past decade has tried to answer some of these preoccupations (including critical thinking, problem solving, democracy, etc.). But the integration of such transversality through traditional disciplines has been really too timid, remaining hampered by a techno-rational and competitive culture in education. Yet, the importance of valuing the role of school as a social actor in order to realize the transforming potential of learning is increasingly recognized (Mezirow, 2009).

ENVIRONMENTAL EDUCATION: AN ONTOLOGICAL REQUIREMENT

Facing this necessity to transform (not only reform) our educational systems, Edgar Morin (2014, p. 11) brings attention to the "anthropological role of education" and insists on its ontological dimension. In this perspective of constructing our being-in-the-world, the ecological dimension of our human identity appears fundamental. Here, environmental education has a major role to play. Its rich theoretical and pedagogical legacy can greatly contribute to the conception and enactment of contemporary education as a process of life, and not awaiting for "real life."

Environmental education—or environment-related education as in French terminology, *éducation relative à l'environnement*—is a core dimension of basic education, which is more specifically centred on one of the three interaction spheres at the basis of personal and social development. Closely related with the sphere of relationship with ourself (learning to learn and to connect with the world, while constructing the multiple aspects of our identity), and with the sphere of relationship with other humans (developing human alterity through democracy, interculturality, peace, justice, cooperation, etc.), environmental education concerns more specifically our relationship with *oikos*, our common home, this living environment where our humanity is connected with the more-than-human world. This third interaction sphere calls for an ecological education: to recognize that we are embodied beings, that our lives are situated and contextualized; to define our human ecological niche in relation with all the niches composing the local and global ecosystem we belong to; to learn how to fulfil this "function" adequately, in a responsible way. It also calls upon an economic education: to learn how to collectively use and share our common home and its resources, with care and solidarity. Ecosophic education is also involved in a transversal way, in order to clarify one's own cosmology (a personal and cultural vision of the world, including our most immediate reality), and to build a coherent ethics, which implies, among other things, rethinking the contextual significance of "responsibility," "justice," "equity," "solidarity," and other socio-ecological values:

> The environment forms, deforms and transforms us, at least as much as we form, deform and transform it. [...] In the space between us and the other (whether person, animal, object, or place ...), each takes on the vital challenge of being in the world. [...] And the crux of human dynamics is in relationship, in ecodependence and in the question of the meaning that each one gives to his existence. (Cottereau, 1999, pp. 11–12)

The environment is neither an object to study or a theme to consider amongst many others, nor is it only an imposed constraint for a "sustainable development." The environment is the web of life itself, at the junction of nature and culture. If education cares for achieving our human being in the world, it has to fully include the sphere of interaction with *oikos*, our environment.

The philosophical and pedagogical field of environmental education has greatly unfolded through the decades and has produced a rich diversity of theoretical and practical currents (Sauvé, 2005). But basically, it aims to forge our ecological identity—our way of being-in-the-world with integrity and integrality—and to stimulate and accompany individual and collective socio-ecological commitment—in the diverse forms it can adopt, often interlinked: philosophical, aesthetic, creative, territorial, political, and others. Commitment is an act of identity: it involves defining ourselves—initially—and to keep on building and clarifying our individual and collective identity. Here, we shall look more specifically at commitment of a political nature, because the environment—as education—is a socially shared object of care, thus of eminently political nature.

ECOPOLITICAL EDUCATION: TOWARDS ECOCITIZENSHIP

Ecocitizenship corresponds to the political dimension of our relationship with the environment. Politics relates (should relate) to our collective involvement in setting the best conditions for living well together, in our common home. The root of the word politics is from the ancient Greek *polis*, which means "city," this democratic place (which should be inclusive) where free and autonomous humans take decisions together about things that affect them all. *Polis* is the "city," the school, the workshop, the neighbourhood, the village, the town, the country, the international community. The idea of politics is then intimately linked to the idea of citizenship: we have to learn to live together in the "city."

Ecocitizenship gives a specific meaning to the "city," that of our *oïkos*, our living place shared with us all humans, and also with all other life forms. In the ecological city, our humanity is intertwined within the fundamental web of life itself: we are part of life systems, we share the same thread of life that links all living beings. The city is not restricted to our human community; it includes the whole community of life.

Since the idea of citizenship is closely linked to the one of democracy, ecocitizenship means enriching democracy with an ecological dimension. Dominique Bourg and Kerry Whiteside (2010) have developed the concept of ecological democracy: "Protecting the biosphere involves rethinking democracy itself" (p. 10). Here, nature is no longer considered secondary, like an element of public affairs to be looked at if there is time and resources to do so. The environment is more than "a place, an historical site, a source of raw materials, a tradable commodity. [...] Nature is an integral part of deliberations within the organization of the city" (pp. 101–102).

Issues regarding hunger, thirst, health, or energy, for example, remind us that human/social realities are closely linked to ecological realities. Murray Bookchin (2005) insists that violence between humans, between societies, and violence against nature have the same roots: they come from the same desire for power and domination, from the same contempt. The notions of socio-ecological equity, ecological justice and more recently, of climatic justice put into light the political dimension of these issues.

Exercising such a democracy requires the development of ecocitizen competencies: mainly critical, ethical, political, and creative competencies (Sauvé, 2015). And the most efficient way to develop these competencies is to immerse oneself reflexively in community projects, social debates, and citizen movements.

Critical competency is based on the capacity to ask questions (what? how? by and for whom?) and demand adequate answers. Paulo Freire (in Freire & Faundez, 1992) associates "learning to question" with a pedagogy of liberation. Ethical competency seeks to answer "why?" What values, what value system, support decisions? What are the values that I/we wish to promote? Here, our relationship with the environment is understood from an in-depth perspective, including its connexions with issues of peace, human rights, solidarity, equity, and other social realities. Political competency develops within the exercise of democratic debate—that must be endlessly claimed as a right. One must learn to occupy the participative and active spaces of democracy efficiently, and contribute to broaden them. Michelangelo Pistoletto (in Morin & Pistoletto, 2015) proposes the word *demopraxis*, which invites us to live a reflexive democracy. Finally, as argued above, if we must learn to denounce and resist, we also have to propose. This is where a creative competency comes into play to imagine solutions, alternative projects. It is focused on divergent thinking and a proactive attitude. It stimulates ecosocial innovation, designing ways of doing things, acting, and being together that contribute to the transformation of our relationship with others and the environment.

The context in which these multiple competencies can be jointly developed is through citizen action and moreover, through ecocitizen action. They feed on commitment and, in return, make it more efficient. Unfortunately, they are rarely taken into account in formal education. Yet, youth can be involved and committed, not as future citizens, but as fully fledged citizens of the ecological city that needs their critical stance, their impetus, their energy, their desire to contribute right now to "real life."

HOW TO PROMOTE AN EDUCATION GROUNDED IN LIFE?

Education for life (ecosocial life), education as a process of life, needs to be supported by sound institutional policies and strategies, as well informal education systems as in non-formal organizations and initiatives—where financial support is dramatically melting away in the neoliberal and conservative societal wave.

Despite the goodwill of last decade educational reforms, the current formal systems still do not promote and support education as life, do not include school in the "city," in the *oïkos*. Teaching–learning situations are trapped in the "right-wing" disciplinary didactic culture imposed by national and international evaluation and competition systems. Of course, in the daily and intimate life of classrooms, courageous, innovative, and socially involved teachers achieve miracles—going against the grain—in order to open up the school to the realities of its environment and to create bridges with other actors of the educational community. It is important to support these initiatives and promote them, particularly through the development of structuring strategies.

In this regard, what about the institutional impulse of UNESCO now referred to in national education policies? Supplanting the previous environmental education international programme (1975–1995), but promoting the same "progressive pedagogy" (interdisciplinarity, critical thinking, project learning, and others) and same institutionalization strategies, the UN Education for Sustainable Development Decade (DESD)—despite its huge politico-economic support—did not succeed in transforming education as suggested by Faure (1972) many years ago, and as observed more recently by Delors (2013). Using virtuous arguments and aiming for the resolution of worldwide problems, the programme imposed (and still does) an economistic worldview—a narrow and distorted cosmovision where economy (seen as growth to be sustained) stands outside society, imposing its rules over the relationship between the environment (as a set of resources not to be exhausted) and the society (as a collection of producers and consumers) (Berryman & Sauvé, 2016; Jickling, 2016). If the principles of sustainable development are supposed to offer some initial ethical framework for business administration and some aspects of public affairs management—arguing that social and ecological issues have to be taken into account so as to better sustain economic growth—there is certainly a huge confusion in considering it as a societal project, and even more as a universally appropriate and holistic education programme.

As well, if the value and language of *sustainability* is often adopted to avoid the hard injunction of sustainable development, it would be important to recognize its minimalist and vague character. Why could (should) sustainability be the core value of our societies, considering the existing spectrum of ecocentric ethics, including those inspired by indigenous cosmovisions, or considering the value of harmony, adopted in the community-centred politics of "*Vivir bien*"?

In its recent proposal *Rethinking Education: Towards a global common good?*, UNESCO (2015) presents a revisited analysis and vision of contemporary education, putting in evidence important concerns. But still, education as a "common good" is presented as "a key resource for the global integrated framework of sustainable development goals" (p. 3). One must recognize that behind the humanitarian discourse of the United Nations Development Project (2015) presenting 17 *Sustainable Development Goals* as a response to problems of poverty, gender inequity, environmental degradation, etc., the global solution proposed remains the promotion of sustained economic growth—thus putting aside the analysis and recognition of the fundamental structural causes of present societal and environmental problems, which are closely connected.

Global Citizenship Education (2014–2021), the international education programme launched by UNESCO in 2014, is also presented as a contribution for achieving the Sustainable Development Goals. Unfortunately, no ecological concerns can be found; there is no place for the development of the ecological dimension of our identity, nor ecojustice in relation with social justice. There is only a timid call for "empathy" for the other humans and the environment—amongst the many objectives.

The texts, declarations, and objectives of UNESCO should be considered as global propositions from which certain elements can be inspiring, but they must not become confining, convergent-thinking moulds, ready-made recipes, or neo-colonization strategies. We must keep examining these top-down guidelines, in the light of various reflective educational fields and taking into account the diversity of territorial, social, and cultural contexts.

For this task, the fields of ecopedagogy, of critical environmental education, of ecocitizenship education, of community education in the context of "Vivir bien" or "Ubuntu," and of other "alter-native" educational theoretical and practical fields, offer important contrasting visions of the world, other diagnosis of contemporary problems, other conceptions of the role of education in societies, and of the identity of educators. They are important because they call for immersion in life, for participation, their conception of "politics" as a collective democratic affair, their ethical reflection that goes beyond anthropocentrism, their recognition of the ontological dimension of education, and their striving for emancipation from any oppression. These pedagogical fields should inspire and nourish educational choices as well at the classroom and community levels as the one of public policies.

And again, in the process of collectively and reflexively weaving education into the fabric of life itself, in this process of living education as a journey of personal and social emancipation, beyond the limits of any exogenous prescription (would it be from prestigious international institutions), let us recognize the "treasure" (in the words of Delors, 2013) to be found in the rich experience of the different actors of our educational society, in the learning dynamics of the people and the groups involved in addressing collective issues in our "city," our *oïkos*. Learning through collective action, reacting against projects or public decisions that are invasive or unjust, or developing ecosocial initiatives that contribute to the transformation or improvement of our way of living here together, is a precious outcome of citizen commitment that should inspire pedagogical practices in school and academic settings.

REFERENCES

Aubenas, F., & Benasayag, M. (2002). *Résister, c'est créer*. Paris: La Découverte.
Baillargeon, N. (2010). *Noam Chomsky. Pour une éducation humaniste*. Paris: Éditions de l'Herne.
Berryman, T., & Sauvé, L. (2016). Ruling relationships in sustainable development and education for sustainable development. *The Journal of Environmental Education, 47*(2), 104–117. doi:10.1080/00958964.2015.1092934
Bookchin, M. (2005). *The ecology of freedom: The emergence and dissolution of hierarchy*. Oakland: AK Press.
Bourg, D., & Whiteside, K. (2010). *Vers une démocratie écologique. Le citoyen, le savant et le politique*. Paris: Le Seuil.
Cottereau, D. (1999). *Chemins de l'imaginaire, pédagogie de l'imaginaire et éducation à l'environnement*. La Caunette, France: Éditions de Babio.
Delors, J. (1996). *L'éducation, un trésor est caché dedans*. Paris: Odile Jacob.
Delors, J. (2013). The treasure within: Learning to know, learning to do, learning to live together and learning to be. What is the value of that treasure 15 years after its publication? *International Review of Education, 59*(3), 319–330.

Faure, E. (1972). *Learning to be. The world of education today and tomorrow.* Paris: UNESCO.

Freire, P., & Faundez, A. (1992). *Learning to question.* New York: The Continuum Publishing Company.

Hayward, B. (2012). *Children, citizenship and environment. Nurturing a democratic imagination in a changing word.* London: Routledge.

Jickling, B. (2016). Losing traction and the art of slip-sliding away: Or, getting over education for sustainable development. *The Journal of Environmental Education,* 47(2), 128–138. doi:10.1080/00958964.2015.1080653

Kincheloe, J. L. (2005). *Critical pedagogy.* New York: Peter Lang.

Manier, B. (2012). *Un million de révolutions tranquilles: Travail, environnement, santé, argent, habitat: Comment les citoyens transforment le monde.* Paris: Les liens qui libèrent.

Mezirow, J. (2009). Transformative learning theory. In J. Mezirow & E. W. Taylor (Eds.), *Transformative learning in practice. Insights from community, workplace, and higher education* (pp. 18–32). San Francisco: Jossey-Bass.

Morin, E. (2014). *Enseigner à vivre.* Paris: Actes Sud.

Morin, E., & Pistoletto, M. (2015). *Impliquons-nous.* Paris: Actes Sud.

Sauvé, L. (2005). Currents in environmental education—Mapping a complex and evolving pedagogical field. *The Canadian Journal of Environmental Education,* 10, 11–37.

Sauvé, L. (2015). The political dimension of environmental education—Edge and vertigo. In F. Kagawa & D. Selby (Eds.), *Sustainability frontiers: Essays from the edges of sustainability education* (pp. 99–112). Farmington Hills, MI: Barbara Budrich Publishers.

Thomashow, M. (1996). *Ecological identity. Becoming a reflective environmentalist.* London: MIT Press.

UNESCO. (2014). *Global citizenship education (2014–2021).* Paris: UNESCO. Retrieved from http://unesdoc.unesco.org/images/0022/002277/227729E.pdf

UNESCO. (2015). *Rethinking education: Towards a global common good?* Paris: UNESCO. Retrieved from http://unesdoc.unesco.org/images/0023/002325/232555e.pdf

United Nations Development Project. (2015). *Transforming our world: The 2030 Agenda for sustainable development.* United Nations. Retrieved from https://sustainabledevelopment.un.org/post2015/transformingourworld

Lucie Sauvé is Professor at the Faculty of Education of the Université du Québec à Montréal (UQAM), Canada, and Director of the *Centr'ERE* research center— Centre de recherche en éducation et formation relatives à l'environnement et à l'écocitoyenneté. She is member of the Institute of Environmental Sciences, the

Health and Society Institute, and Dialog—Québec Research Network on Indigenous issues. Professor Sauvé is the founder and co-responsible of the UQAM postgraduate programme in environmental education. She is co-founder and Director of the international research journal *Éducation relative à l'environnement – Regards, Recherches, Réflexions*. Lucie Sauvé has an extensive experience with international cooperation projects in Latin America (Amazonian region) and won, in this context, two excellence awards. Her current research focuses on environmental health education (agriculture, food, and water issues), energy education, ecocitizenship education, and environmental educators' and community leaders' professional development. In 2009, she was Co-president with Bob Jickling of the 5th World Environmental Education Congress and received an award from the ADEREQ— Association des doyens et directeurs pour l'enseignement et la recherche en éducation du Québec —as a recognition of the excellence of her contribution to research in education. In 2015, she received a *honoris causa* doctorate from the Universidad Veracruzana (Mexico) for her academic and social involvement. www.centrere.uqam.ca

Resilient Education: Confronting Perplexity and Uncertainty

Edgar J. González-Gaudiano and José Gutiérrez-Pérez

Abstract The chapter explores the role that education plays in a context of growing conflict that magnifies the usual challenges faced by environmental education. The promotion of extractive megaprojects (opencast mining, micro-dams, shale gas extraction), including several preconized as alternative production and sustainable energy strategies (giant wind turbines), has resulted in social conflicts as a result of the breakdown of community ties, the destruction of regional economies, the loss of cultural diversity and the degradation of environments. In areas where such investments are located, local relationships have been disjointed and then selectively integrated and subordinated to globalized value chains led by large transnational corporations. This chapter ends with consideration of strategies that can be undertaken to strengthen local resilience against the onslaught of huge economic forces that tend to elicit the subjection of local governments.

E.J. González-Gaudiano (✉)
Institute of Researches in Education, Universidad Veracruzana, Xalapa, Mexico
e-mail: egonzalezgaudiano@gmail.com

J. Gutiérrez-Pérez
Educational Research Methodology Department, University of Granada, Granada, Spain
e-mail: jguti@ugr.es

© The Author(s) 2017   125
B. Jickling, S. Sterling (eds.), *Post-Sustainability and Environmental Education*, Palgrave Studies in Education and the Environment,
DOI 10.1007/978-3-319-51322-5_9

Keywords Environmental education · Neoextractivism · Resilient education

It has been difficult to find an appropriate way to begin this chapter. We decided to approach resilient education as a pedagogical and political dimension of education that we consider most coherent with the other sections of this book. This view provides critical elements to combat perplexity and uncertainty in situations of inequity, inequality, and improper exploitation of natural resources. Resilient education represents an emancipatory tool to develop critical thinking and active involvement of citizens in the problems of their environment. The aim of this book as a whole affords us the opportunity to glimpse a horizon of possibility for the near future.[1] As Harari (2013) points out, a "horizon of possibility" means the entire spectrum of beliefs, practices, and experience that are open before a particular society, given its ecological, technological, and cultural limitations.

After a decade of education for sustainable development, and in the light of persistent financial and economic crisis, we are now witnessing various phenomena that did not form part of the *Zeitgeist*, or spirit of the era, that pervaded a period of 40 years of environmental education around the world hitherto. On the one hand, the phenomena of environmental degradation have been magnified to unimaginable scales since the beginning of the new millennium (Worldwatch Institute, 2015; IPCC, 2014). New extraction techniques to meet the demands of an energy model in crisis, as well as the materials required to reactivate economic processes, have given rise to unprecedented neocolonialist expressions (Machado, 2013). On the other hand, information and communications technologies have rapidly given rise to multiple social processes that, at the same time, numb critical consciousness through aggressive marketing and subdue identity through the siren song of consumption (Plepys, 2002). This numbing is coupled with, among other factors, the resurgence of hotspots fraught with extremist ideologies that many forecasts suggest will lead to new scenarios of polarization and armed conflict, terrorism, and chaos (Kegley & Blanton, 2016).

In this perplexing context, educational systems barely react with the timing and depth required. For example, regardless of what is happening in other spheres of education, school curriculum is depleted; that is, it has lost its heuristic capacity to explain make sense of what happens outside the classroom (González-Gaudiano, 2007). In the two countries represented by this chapter's authors, education for sustainable development has not been established as a public policy. An unfortunate consequence is that programmes and centres of environmental education, that had constituted important achievements at the end of the century, are gradually being dismantled. Through crisis and financial pressure, precarious advances in this field have become dispensable.

Given such a scenario, we have decided to focus on those educational processes that can provide us with means to recover the dynamics of our lives in order to combat the ravages of what are, in every sense, difficult times. In other words, we promote resilient education that, through a socio-critical, emancipatory, and political view, focuses above all, on the skills that will be increasingly required to tackle the tremendous environmental degradation and climate change that is occurring in the world. This is especially a concern in the so-called developing world where the living conditions and comforts come at a very high environmental and social cost.

It is no longer sufficient to promote critical education. We must develop resilient educational activities through awareness of the abuses, risks, outrages, limits, and atrocities that the present has thrown upon us; an education armed with criticism and charged with positive energy to actively overcome injustices, traumas, and growing inequalities; an education to combat the increasing vulnerability of our lives; a resilient education informed by a sense of tragic optimism (Santos, 2009; González-Gaudiano, 2016).

THE SITUATION AHEAD OF US

When environmental education appeared in the early 1970s, serious concerns existed regarding the deterioration of the environment. At that time, these concerns stemmed mainly from the problems of pollution deriving from urban expansion and industrialization processes. Without examining these aspects in depth in this chapter, we can point to a number of authors who denounced this situation in books that acquired cult status among the emerging environmentalist movement (i.e. Rachel Carson; Barry Commoner; E.F Schumacher). In response, and based on previous experiences, the first environmental education programmes were developed, although with hindsight their scope can now be described as limited.

From an educational perspective, the problem we now face has taken on a magnitude and complexity that exceeds our previous worst-case scenarios. The neoextractivist model, that characterizes the current "style of development" (Svampa, 2012), is seen as the production process that combines private capital—normally transnational—with state participation, where the latter provides different tax incentives, flexible labour regulations, and environmental deregulation. The dynamic core of the model is based on technologies that move large volumes of unprocessed, or only partially processed, materials for export as "commodities" to the international market. The consequences include enormous socio-environmental deterioration and territorial fragmentation, with marginalized areas and extractive enclaves associated with global markets (Gudynas, 2009). Exploitation processes are concentrated especially in oil and shale gas production through hydraulic fracturing (fracking), opencast mining, and intensive agro-industrial production in the form of, for example, transgenic monoculture and biofuel production. Other interventions include infrastructure projects promoted as alternative energies with low carbon emissions, such as hydroelectric dams (large and micro) and wind power projects.

The distinctiveness of this renewed model of capitalist development, relative to conventional exploitation practices, is expressed through three characteristics that have gained importance in recent decades: (a) overexploitation of increasingly scarce natural resources, (b) the expansion of extractive boundaries towards territories previously considered "unproductive," and (c) the tendency towards monoculture associated with extensive farming practices (Svampa, 2011). The model also induces subordination and dependence as extraction megaprojects are integrated in "global value chains" that control many production and marketing processes. Additionally, this model is based on neocolonial regimes, often with ambiguous legitimacy and low quality of governance at the local level.

The convergence of the global and local dimensions gives rise to "territoriality tensions" (Porto Gonçalves, 2001) created through antagonistic processes of social organization and asymmetric forces. Establishing extraction megaprojects entails the complete reconfiguration of the territories in which such projects are developed. Social, productive, ecological, political, and even educational relationships are subsumed by an instrumental logic that leads to the breakdown of community ties, the destruction of regional economies, the loss of cultural diversity, and the degradation of environmental conditions.

Thus, the territories where these seemingly progressive projects are located have been fragmented and disjointed from the processes and flows of local relationships. Even more, these living spaces are undertaken selectively and subject to globalized value chains led by large transnational corporations (Machado, 2009).

Curriculum development is also tied to this process of instrumental sublimation through hegemonic actions. Learning skills are homologated and certified through implementation of standardized competences that enable the tyranny of inter- and intra-territorial comparability (Apple, 2004). This approach to curricula ignores the uniqueness of contexts and horizons of possibility (in the Harari sense, 2013) that are inspired by the spectrum of beliefs, practices, imaginaries, and expectations of social change of the different sectors within the same territory.

However, despite being presented as an effective development strategy to avert the scourge of poverty, the neoextractive model, and the resulting "territorial alienation" (Santos, 1996), faces massive and radical resistance at local levels from organized solidarity networks (grass roots). With the aid of information technologies, this resistance has spread to the continental level, giving rise to a "new internationalism." These movements revolve around the recovery of the "commons," the "good life," and "living well" (Gudynas y Acosta, 2011), and "living together" (Hatzfeld, 2011) in cities, with their own vindicating interpretation of a "social spring" and with nuanced resignification specific to each territorial context. This provides insight into hitherto hidden, unheard, or silenced suffering of the territories' peoples. Likewise, it reaffirms the legitimacy of the demands they pursue, and it strengthens the arguments in favour of the struggles of the "common people" (Sauvé, 2013).

The emphasis of these struggles in defence of "commons" turns them into what Harvey (2003) described as "insurgent movements against accumulation by dispossession." Svampa (2008) presents them as "socio-environmental movements" considering that they emphasize, in the deepest sense, an integral relationship with the environment—that is, vital space cannot be separated from the social world. While mobilized communities struggle to maintain their traditional ways of life by defending the commons of nature, it is not *a priori* a political option. Rather, it is a defensive reaction against predatory coercion that gradually embraces other collective and symbolic characteristics. Broken biographies like those of Chico Mendes in Brazil and Berta Cáceres[2] in Honduras lead this zealous and legitimate vindication tradition to defend what is common and denounce larceny on the part of multinationals protected by local governments. The boomerang effect of the exploitation of populations, the contempt for life, and the legalized abuse of collective resources is painfully felt, and outrage is increasing dramatically (Sauvé, 2013).

There are countless examples of the neoextractive model at work in Latin America (abandoned phosphate ponds near communities in Chile, the exile of Indigenous peoples in the Brazilian forests for timber extraction or water privatization wars). The same painful outcomes are repeated in Asia (dispossession and persecution of Maoist peasants in India, inhuman coal mining in China) and Africa (children forced to leave school early to extract gold and uranium which is then smuggled into Burkina Faso, Niger, and Namibia). Arundhati Roy (2015) referred to all these people as the "ghosts of capitalism."

Refined examples of how the tigers of Asian, European, or American monopolies extend their claws to every corner of the planet are vividly exemplified under the guise of corporate philanthropy and the implementation of social responsibility—that is, wolves in sheep's clothing. The Rockefeller and Ford Foundations in India, for example, provide generous donations to universities, fellowships for artists, filmmakers, and activists enacting purposes such as "reorienting the future of humanity."[3] By intervening in the dynamics of popular movements, their legitimate daily interests and desires are perverted and redirected towards the "consumerism dream." In another tragic example, the model of microcredits for starving farmers promoted by Yunus and the Grameen Bank in India has caused a large number of suicides among women trapped by this change in their "existential horizon" (Kinetz, 2010). That is, their deep despair is due to high interest rates that micro-lenders charge their clients, and their coercive loan recovery tactics.

It is clear that environmental education faces new challenges that must be tackled using qualitatively different strategies, from the pedagogical and political standpoint. This will neither be easy nor will it yield immediate results, but we feel it must be addressed by adopting an attitude of "tragic optimism." Here, tragic optimism is a combination of an acute awareness of the struggles for emancipation and the difficulties of resistance to co-option coupled with an unshakable confidence in the human capacity to overcome seemingly insurmountable difficulties and to create potentially infinite horizons of possibility (Santos, 2009; González-Gaudiano, 2016).

Environmental educators are particularly challenged by hunger, thirst, fear, oil "spills," toxic mining, the ravages of climate change—and all tragedies that occur, and multiply. They reveal more than ever the extent to which ecological and social problems are inseparable. These problems are entangled in a network of interactions (Sauvé, 2013) that clearly show the "metabolic breakdown" between capitalism and the planet (Foster, 1999).

In the meantime, the World Economic Forum (2015) insists on an agenda of debates on artificial intelligence, the future of health, or action against climate change. The events that determine its agenda, and road-map, always fall outside of the perimeter of action. So why then does this forum convene, and make pointless gestures on matters it claims to be of urgent concern, unless they are merely showcases to divert attention from the core of decisions and problems?

TOWARDS A DIFFERENT ENVIRONMENTAL EDUCATION

Environmental education cannot be alienated from its surroundings and build social theorems in a vacuum. Nor can educators develop aseptic, neutral, and timeless interventions without taking into account the political, social, economic, philosophical, and scientific circumstances of each moment in history. Environmental education's actions, as a sphere of scientific, technical, and artistic knowledge, and as a disciplinary spring-board, are subject to the constraints of their space and time. It is unavoidably a heuristic of present, local, and global circumstances.

All education must be environmental; otherwise it is not education. All initiatives in the field of education are dependent on near and distant contextual variables. Given their synchronous nature, such initiatives must act as catalysts to reveal economic, social, environmental, and—ultimately—human contradictions in the frame of their historical moments.

Going forward, our work is based on theoretical postcolonial architectures developed by Bauman (2015), Latouche (2008), Piketty (2014), and Santos (2009); on societies in transition from solid to more liquid, virtual, and volatile states; on the results of the fast-paced transformations of our time; and are subject to a transfer of ephemeral values and existential illusions. We now propose to analyse how this theoretical discourse has influenced the renewal of the aims and strategies of environmental education. We also ask ourselves what professional challenges the new scenario poses for environmental educators in the twenty-first century. And, if schools are the main educational context that legitimize and formalize their own identities as a useful and necessary socio-professional institutions, then what tools and skills should they use to shed light on, and understand, the motives that have driven certain tricksters to: establish anti-values that reward profitmaking, speculation, and monetary enrichment, as if this were a "noble end;" or to promote causes that are prioritized ahead of the most basic human rights such as housing, work, the right to free higher education, a dignified existence, and harmonious respect for natural ecosystems?

In the context of these ferocious changes and unstoppable transformations, environmental educators can have difficulty keeping pace with, and understanding real events. It is also a challenge to update their working methods and professional skills in order to provide comprehensive and effective responses to emerging challenges. Muñoz Molina (2013) offers a brief summary of the solutions:

- We need accurate information about changes and evolving conditions in order to hold rational opinions, and to know which mistakes need to be corrected and which achievements can be used to find solutions in this emergency.
- We urgently need to understand the rapid changes that are happening around us, and since there can be no understanding without words; we need these words to be as clear and precise as possible.
- A peaceful civic rebellion is required which, like the American civil rights movement, makes intelligent and astute use of all the mechanisms provided by law and the full force of mobilization to rescue the sovereign territories usurped by the political class.
- The need to take advantage of everything has often been replaced by the capricious habit of wasting everything. It is unforgivable that water and other basic resources are being wasted when many millions of people in the world cannot enjoy them.
- Everyone almost always has the power to do something good or badly, to be rude or polite, to throw a crumpled bag, a bottle or a can of soda on the ground or place it in a dustbin, to shout or whisper, to cry out in anger at criticism, or stop and check if criticism is fair. These powers need to be consciously exercised.

Bauman (2005, p. 12) mentions a number of trends that have accompanied civilization since time immemorial. For example, a habitual response to "wrong behaviours," "unsuitable conduct," and "undesirable outcomes" is education, or re-education. The thinking has been, according to Bauman, that by instilling in learners new motives, propensities, and skills they will be able to fight back, and challenge the impact of daily experience. These habitual responses, despite being repeatedly refuted as ineffective by the realities of social life, have been clung to stubbornly in the hope of "using education as a jack potent enough to unsettle and ultimately to dislodge the pressures of social facts...education in such cases challenge the impact of daily experience, to fight back and in the end defy the pressures arising from the social setting in which the learners operate" (p. 12).

Rather, we contend that resilient education can provide solid responses to emerging challenges if educational curricula are seen as a set of reasons, interests, and controversial arguments that apply pressure and stimulate continued criticism in modern society (Jickling and Wals, 2008; Popkewitz, 2009). From this perspective, social and scientific development, innovation, progress and human dignity, the challenges for educators in general and environmental educators in particular, are unlimited. Environmental education is possibly the optimal field for experimenting with proposals that respond to challenges arising from our current historical context.

Assuming that non-formal organizations and educational systems at different levels can transmit and reproduce a calculated selection of content and cultural values (McLaren and Kincheloe, 2008), it is valid to question: the principles of models of environmental literacy and curriculum greening that are currently promoted (Kemmis and Mutton, 2012); different types of environmental education resources, centres and programmes; and, guidelines foisted upon educators by international institutions that attempt to force us to accept non-neutral languages, such as "education for sustainable development." At the end of the day, the "international" vision of development is imposed and integration is demanded. Educators are expected to transfer ideas of education for sustainable development to their programmes, activities, and methodologies. This results in a type of neutral and seemingly innocent education that must not include controversial issues, challenges to contemporary economics, or political analyses of social conflicts that give rise to the exploitation of resources.

Educators and citizens living in different territories are somewhat bewildered and worried about the many changes looming on the horizon that inevitably affect the conditions, hopes, and expectations that guide their work. In the present era, there are very few certainties in our lives and many uncertainties on several fronts of economic, ecological, and political life. Nevertheless, we highlight two of these uncertainties because we believe they may directly condition the way we understand resilient education and related curricular models. To this end, consider:

- That the ideology of economic growth and progress is a fallacy based on revolving demands of global interests that feed on false premises; and, in implementing this ideology, citizens are placed in a state of collective hypnosis while root causes are concealed. That we have to learn to live differently, and school curricula must teach to build new models of coexistence seated in a citizenship able to critically analyse the errors of our aspirations in becoming a biocentric species, yet falling short through self-absorbed, short-sighted, and ill-advised ethics.
- The false premises that circulate around us represent serious obstacles to a real transition to a more sustainable world. They include some of the most widespread beliefs that underpin the "culture" of productivist/consumerist society (Riechmann, 2008). Most even contradict the basic laws of thermodynamics, namely that the economy can grow indefinitely in a finite biosphere. Further false premises include that competitive markets allow us to efficiently cope with shortages; and technology, in practical terms, makes us omnipotent.

The changing nature of everyday life and the biophysical, social, and political world has been a constant throughout the history of humankind. However, the current pattern of global change is unprecedented, exhibiting characteristics that make it different from previous historical periods (González & Montes, 2010). These include extremely rapid, intense, and global-scaled changes; impacts and consequences in all natural systems at all scales; failure of natural and social systems to co-evolve with the speed of change; and, uncertainty and unpredictability in the contemporary world.

The innovations and surprises that each era brings are coupled with disciplinary changes, conceptual tensions, and advances in systematic research (Tuck & McKenzie, 2015) based on a new resilient epistemology. Contemporary education needs resources capable of meeting these new social demands. Educators need to be equipped with a resilient "microchip" to tune real frequencies and voltages to their response capacity, providing them with: enough energy to meet the challenges with tragic optimism, to successfully respond to the circumstances in the critical reconfiguration of changing realities, and to foster development of skills in new generations. In summary, they need to rebuild their reality on an ongoing basis and address uncertainty in an informed way.

Resilient education is a tool for social empowerment that cannot ignore relationships between science and power (the power of citizens). Importantly, citizens are not beholden to so-called "value-neutral" science. A new epistemology of resilience should enable them to negotiate, resist, denounce, convince, decide, create, and transform. By encouraging critical spirit through a search for valid information, resilient education participates in the exercise of enabling an enlightened democracy. As Sauvé (2010) argues, citizens must, in a short time and bereft of means, try to understand extremely complex issues, find and process huge amounts of information, relate and critically interpret this information, and synthesize it in an integrated and meaningful manner. In this sense, resilient education organizes and gives a sense of reality to the systemic contributions of the science of sustainability and it can help establish broad intervention strategies beyond the here and now.

NOTES

1. The concept of resilience in the social sciences has been useful to characterize a set of phenomena related to individual or community capacities to bounce back from adverse situations of disturbance and instability and to recover conditions of self-organization (Adger, 2000). This chapter emphasizes a meaning of the concept that focuses on resilience from a political, ecological, and educational perspective. We are mindful, however, that the use of "resilience" is also currently being contested (Neocleous, 2013).
2. Assassinated in early 2016, Berta Cáceres led a movement that in 2013 and in 2014 got the World Bank and the multinational Chinese Sinohydro, to desist from building a hydroelectric dam on the Gualcarque River, listed by the Lenca Indians as sacred and crucial for their survival.
3. Through foundations and philanthropies, elite groups and corporations exert influence in decision-making at different political and social levels. Examples include the Magsaysay award for community leaders from Asia, and other funding of certain social movements and leaders (Negi, 2008).

REFERENCES

Adger, W. N. (2000). Social and ecological resilience: Are they related? *Progress in Human Geography*, 24(3), 347–364.

Apple, M. W. (2004). *Ideology and curriculum*. London: Routledge.

Bauman, Z. (2005). *Liquid life*. Cambridge, MA: Polity Press.

Bauman, Z. (2015). *Modernidad líquida*. México: Fondo de Cultura Económica.

Foster, J. B. (1999). Marx's theory of metabolic rift: Classical foundations for environmental sociology. *American Journal of Sociology*, *15*(2), 366–405.

González, J. A., & Montes, C. (2010). Cooperación para el desarrollo en tiempos de cambio global: cuando seguir haciendo lo mismo ya no es una opción. Fundación IPADE: *Cuatro grandes retos, una solución global* (pp. 8–25). Madrid: Fundación IPADE por un desarrollo sostenible.

González-Gaudiano, E. (2007). Schooling and environment in Latin America in the third millennium. *Environmental Education Research*, *13*(2), 155–169.

González-Gaudiano, E. J. (2016). ESD: Power, politics, and policy: "Tragic optimism" from Latin America. *The Journal of Environmental Education*, *47*(2), 118–127.

Gudynas, E. (2009). Diez tesis urgentes sobre el nuevo extractivismo. Contextos y demandas bajo el progresismo sudamericano actual. En VVAA. *Extractivismo, política y sociedad* (pp. 187–225). Quito: CAAP/CLAES.

Gudynas, E., & Acosta, A. (2011). El buen vivir o la disolución de la idea del progreso. In M. Rojas (coord.). *La medición del progreso y el bienestar. Propuesta desde América Latina* (pp. 103–110). México: Foro Consultivo Científico y Tecnológico.

Harari, Y. N. (2013). *From animals into gods: A brief history of humankind*. Kinneret, Israel: Dvir Publishing House Ltd.

Harvey, D. (2003). *The new imperialism*. Oxford, UK: Oxford University Press. Retrieved from http://eatonak.org/IPE501/downloads/files/New%20Imperialism.pdf

Hatzfeld, H. (2011). *Les légitimés ordinaires. Au nom de quoi devrions-nous nous tarie?* Paris: L'Harmattan.

IPCC. (2014). *Climate change synthesis report 2014: Summary for policymakers*. Retrieved from http://www.ipcc.ch/pdf/assessment-report/ar5/syr/AR5_SYR_FINAL_SPM.pdf

Jickling, B., & Wals, A. (2008). Globalization and environmental education: Looking beyond sustainable development. *Journal of Curriculum Studies*, *40*(1), 1–21.

Kegley, C., & Blanton, S. (2016). *World politics: Trend and transformation, 2016–2017*. Toronto: Nelson Education.

Kemmis, S., & Mutton, R. (2012). Education for sustainability (EfS): Practice and practice architectures. *Environmental Education Research*, *18*(2), 187–207.

Kinetz, E. (2010, October 19). Suicides spark scrutiny of Indian microfinance. *Bloomberg BusinessWeek*. Retrieved from http://www.microfinancetransparency.com/evidence/PDF/12.26%20Business%20Week%20article%20India%20suicides.pdf

Latouche, S. (2008). *La apuesta por el decrecimiento. ¿Cómo salir del imaginario dominante?* Barcelona: Icaria.

Machado, H. (2009). Minería transnacional, conflictos socioterritoriales y nuevas dinámicas expropiatorias. El caso de Minera Alumbrera. En M. Svampa y M.

Antonelli (eds.) *Minería transnacional, narrativas del desarrollo y resistencias sociales* (pp. 205–228). Buenos Aires: Biblos.

Machado, H. (2013). Crisis ecológica, conflictos socioambientales y orden neocolonial: Las paradojas de NuestrAmérica en las fronteras del extractivismo. *Revista de Estudios Latino americanos (REBELA), 3*(1), 118–155. Retrieved from http://rebela.emnuvens.com.br/pc/article/view/137/269

McLaren, P., & Kincheloe, J. L. (2008). *Pedagogía crítica: De qué hablamos, dónde estamos.* Barcelona: Graó.

Muñoz Molina, A. (2013). *Todo lo que era sólido.* Barcelona: Seix Barral.

Negi, R. S. (2008). Magsaysay award: Asian Nobel, not so noble. *Economic & Political Weekly, 43*(34), 14–16.

Neocleous, M. (2013, March/April). Resisting resilience. *Radical Philosophy, 178*, 2–7.

Piketty, T. (2014). *Capital in the 21st century.* Cambridge, MA: Harvard University Press.

Plepys, A. (2002). The grey side of ICT. *Environmental Impact Assessment Review, 22*(5), 509–523.

Popkewitz, T. (2009). Curriculum study, curriculum history, and curriculum theory: The reason of reason. *Journal of Curriculum Studies, 41*(3), 301–319.

Porto Gonçalves, C. (2001). *Geo-grafía:. Movimientos sociales, nuevas territorialidades y sustentabilidad.* México, DF: Siglo XXI.

Riechmann, J. ((coord.)). (2008). *¿En qué estamos fallando? Cambio social para ecologizar el mundo.* Barcelona: Icaria-Antrazyt

Roy, A. (2015). *Espectros del capitalismo.* Madrid: Capitán Swing Libros.

Santos, B. S. (2009). *Una epistemología del sur: La reinvención del conocimiento y la emancipación social.* Buenos Aires: CLACSO-Siglo XXI.

Santos, M. (1996). *Metamorfosis del espacio habitado.* Barcelona: Oikos-Tau.

Sauvé, L. (2010). Educación científica y educación ambiental: Un cruce fecundo. *Enseñanza de las Ciencias, 28*(1), 5–18.

Sauvé, L. (2013). La educación ambiental y la primavera social. *Jandiekua. Revista Mexicana de Educación Ambiental, 1*, 31–42.

Svampa, M. (2008). *Cambio de época. Movimientos sociales y poder político.* Buenos Aires: Siglo XXI.

Svampa, M. (2011). Minería y Neoextractivismo Latinoamericano. Retrieved from http://www.dariovive.org/?cat=12

Svampa, M. (2012). Movimientos sociales, gobiernos y nuevos escenarios de conflicto en América Latina. En C. Moreira, & Avaro, D. (coords). *América Latina hoy. Sociedad y política* (pp. 15–68). Buenos Aires: Teseo, UABC, CPES, FEyRI.

Tuck, E., & McKenzie, M. (2015). Relational validity and the "where" of inquiry place and land in qualitative research. *Qualitative Inquiry, 21*(7), 633–638.

World Economic Forum. (2015). *Annual meeting.* Retrieved from https://www.weforum.org/events/world-economic-forum-annual-meeting-2015/
Worldwatch Institute. (2015). *State of the world 2015: Confronting hidden threats to sustainability.* Washington, DC: Covelo & London, Island Press.

Edgar J. González-Gaudiano is Senior Research Fellow at the Institute of Researches in Education, at the Universidad Veracruzana, Mexico. Gonzalez-Gaudiano is a member of the Commission of Education and Communication of IUCN and former Regional President for Mesoamerica (2001–2006). He has been member of Board of Directors at the North American Association for Environmental Education (NAAEE) (1992–1997). He is a Latin American liaison person and was a member of the UNESCO Group of Reference for the UN Decade of Education for Sustainable Development (2005–2014). He has received a number of awards: the National Award to Ecological Merit (2004) and the UANL Award of Research in Humanities (2007). His current research focus is on social representations of climate change, vulnerability, risk and social resilience.

José Gutiérrez-Pérez is Professor at the Educational Research Methodology Department, Faculty of Sciences Education in University of Granada (Spain), where he teaches Program Evaluation and Research Methodologies in Environmental Education. He is Director of PGC of Training Secondary Teachers and Environmental Education and head of Research Group: "Evaluation in Environmental Education, Social and Institutional" and was awarded in 2004 with the Second National Prize for Educational Research. He played a key role in the development of the Spanish National Strategy for Environmental Education. He has published a number of books and is a member of the editorial board of several international environmental education journals. His research focuses upon the quality criteria evaluation of environmental education centres, educational programmes, and climate literacy.

An Afterword

Stephen Sterling and Bob Jickling

Abstract In this Afterword, the editors reflect on some of the main themes emerging from the collected chapters. They note that the UNESCO Global Monitoring Report, *Education for People and Planet*, was launched just as the first draft of this book was completed. This report is welcomed as it couples the status of education with planetary prospects. Yet, it also serves to underscore the very purpose of this book by failing to recognize the nature and depth of change required in educational practice to meet the aspirations of the report's subtitle, "creating sustainable futures for all." Rather, as reflected by the different authors in this book, the editors suggest that there needs to be a disruption of dominant assumptions in educational thinking and purpose so that a cultural shift towards practice that is life-affirming, relational, and truly transformational can take root. This can be realized at any level of engagement through the role of "rebel teacher" and through "being differently" in the world.

S. Sterling (✉)
Centre for Sustainable Futures (CSF), University of Plymouth, Plymouth, UK
e-mail: stephen.sterling@plymouth.ac.uk

B. Jickling
Lakehead University, Thunder Bay, Ontario, Canada
e-mail: bob.jickling@lakeheadu.ca

© The Author(s) 2017   139
B. Jickling, S. Sterling (eds.), *Post-Sustainability and Environmental Education*, Palgrave Studies in Education and the Environment,
DOI 10.1007/978-3-319-51322-5_10

Keywords Education · UNESCO · Life-affirming · Relational ·
Transformational · Rebel teacher · Being in the world

At about the time that we submitted the manuscript for this book,
UNESCO published its Global Education Monitoring Report summary
titled, *Education for People and Planet: Creating Sustainable Futures for
All*. As a high-level and influential document for policymakers, it is significant that it links education directly to a sustainability agenda. But some of its
key assertions bear scrutiny in relation to the theme of remaking education,
as represented in this book. Consider, for example, the words of Irina
Bokova, the Director General of UNESCO in the Foreword to the report:

> we must fundamentally change the way we think about education and its role
> in human well-being and global development. Now, more than ever, educa-
> tion has a responsibility to foster the right type of skills, attitudes and behavior
> that will lead to sustainable and inclusive growth. (UNESCO, 2016, p. 5)

While laudable, the idea of fundamentally changing the way we think
about education is thwarted by her following sentence. Zygmunt
Bauman (see González-Gaudiano and Gutiérrez-Pérez) takes us to the
heart of this problem when he says:

> A habitual answer given to the wrong kind of behaviour, to conduct unsui-
> table for an accepted purpose or leading to undesirable outcomes, is educa-
> tion or re-education: instilling in the learners new kinds of motives,
> developing different propensities and training them in deploying new skills.
> (2005, p. 12)

Bauman then challenges the efficacy of this strategy and goes on to
observe:

> The hopes of using education as a jack potent enough to unsettle and
> ultimately to dislodge the pressures of "social facts" seem to be as immortal
> as they are vulnerable (p. 12)

Indeed. This simple aim of instilling motives, developing propensities, and
deploying new skills has never taken educators deeply into important
educational questions.

Bokova's prescription for education underscores just how difficult making fundamental change in education will be. By insisting that education's aim is still framed in the language of "growth" also strikes us as self-defeating. As Sterling observes in this book, UNESCO has been suggesting the need for a "new vision" of education for some time yet what is often missing is a sufficient critique of the dominant cultural worldview which would then help clarify the grounding of necessary alternatives and possibilities (see also González-Gaudiano and Gutiérrez-Pérez; Sauvé). In practical terms, Bauman's key question about educators takes us to a critical point. He asks, "Will they themselves be able to avoid being enlisted in the service of the self-same pressures they are meant to defy?" (2005, p. 12).

This brief critique serves to bring the aims of this book into focus. All authors are, in their own ways, seeking to fundamentally disrupt dominant visions of education and its role in human well-being and global development, and to propose grounds for necessary change appropriate to the global challenges we now face. As the Earth is rapidly shifting from the Holocene to something being called the Anthropocene, there is a collective urgency amongst us to "educate a generation of students who grow dangerous to the status quo" (Orr, in preface to this volume).

Remaking Education and the Rebel Teacher

Talk of quality education, transformative learning, and education as a common good is ubiquitous in international discourse. Yet, there is seldom any in-depth guidance about what is meant by these terms, or how they might be realised. Indeed, transformation is harder than it may seem, often painful, and ultimately visible only in hindsight (Lotz-Sisitka; Jickling). Frequently, responses to anthropogenic environmental degradation have taken the form of some sort of "bolt-on addition to a super-ordinate pre-existing educational structure whose basic motivations were elaborated without any reference to ideas of sustainability" (Bonnett). In this kind of manifestation environmental, sustainability, and sustainable development education can only expect to have limited impact. Seen another way, however, these educations can also offer openings—invitations to look deeply into what is fundamentally wrong with education (Jickling). Or as Sterling says, "environmental and sustainability education have never been, and cannot be, ends in themselves, contained and complete. Rather, they imply and act as outriders or a vanguard for a necessary deeper shift in educational culture." The task at hand is, in part, to revision what education can be, but more emphatically, it is to begin the concrete task of remaking education and the educational culture in which it sits. This work is to be done by "real people who are immersed in real places" (Blenkinsop and Morse).

These real people working in real places will inevitably be "rebel teachers." In the face of ever-present pressures to conform to the status quo—to an increasingly standardised, narrowly conceived, and internationally imposed framing of education—their task will be rebellious. When faced with seemingly overwhelming forces, the only choice will be for everyone to find, in their own way, something to do in response to the challenges. Rebel teachers will need to be ever-vigilant and prepared to name, resist, and denounce those things that they cannot accept and to propose and enact those visions for education that they can say yes to (Blenkinsop and Morse; Sauvé). This book exists to help readers find openings for themselves, and to encourage them to take part in the task of remaking education.

It is important to acknowledge that the authors in this book are, themselves, real people working in real places—and in their own ways they are rebel teachers. It would be incorrect to say that they are armchair scholars. They are not simply writing about abstractions, rather they are writing from decades of concrete experiences as educators and researchers working in real social settings. In important ways their writing reflects their own experiments in remaking education.

On Being in the World

We invite you to consider an additional theme that runs throughout the book—that is, being in the world. It is not enough for any of these authors to imagine bringing about change by simply thinking about the world differently, it is critical for people to be in the world differently, too. A number of years ago, Jim Cheney and Anthony Weston (1999) provided an engaging oeuvre into this kind of thinking. They maintained that how we carry ourselves in the world—that is the etiquette we travel with—will determine what we come to know about the world. They called this knowing "ethics-based epistemology." In other words, our relationship with the world—or our way of being in the world—comes first and shapes what follows. This kind of thinking, which carries the promise of agency and inspiration, resonates deeply throughout the book.

Some authors discuss being and becoming in the world in relation to small-scale teaching and research experiments. For Lotz-Sisitka, this shifts the focus of education from enculturation towards aspiration and describes the outcomes as a kind of learning activism. Sauvé suggests that resistance can become "an act of creation, able to produce itself through other values." And, like Lotz-Sisitka, González-Gaudiano and Gutiérrez-Pérez focus on recovery of the "commons" and "commoning" activities. And they advocate carrying an attitude of "tragic optimism" into the world. By this they mean holding together an acute awareness of the difficulties of emancipation and resistance at the same time as an unshakable confidence in the human capacity to overcome seemingly insurmountable difficulties.

Other authors discuss being in the world as necessarily relational. For Bonnett, "if we are attentive, our being becomes infused with . . . interplay of anticipations. Without it, we enter ontological freefall: our lives untouched and unsustained by a world that we pass through but do not inhabit." For him, authentic education needs to be grounded in authentic human being, fully and bodily engaged with nature; it gives primacy to the ontological over the epistemological.

Interestingly, Le Grange acknowledges, like many other authors in this collection, that interpolating sustainability into mainstream discourse has done little to arrest deepening poverty or the erosion of the world's biophysical base. In response, he first gives readers options to rethink sustainability, but then, intriguingly, speculates about ways of transcending this term entirely. Like Bonnett, he sees this transcendence as relational and ontological. His ontology of immanence can arise when human thought is bent by the earth/cosmos and the individual self dissolves into relationship with the world. Here the subjective being in the world becomes ecological and the imperative to care arises within us as part of our ecological being in the world.

There are many possible ways of framing discussion about themes in this book. We have presented just a few. We hope the organisation of chapters will help readers find those most immediately interesting, but also those themes that emerge as new and challenging. It is, however, a flawed process. Each chapter reflects multiple themes that weave in and out other chapters. The book is a tapestry of ideas that defies simplified organisation. So, in the spirit of the book's intent as a heuristic—or agent of discovery—we invite readers to trace their own interests and priorities through the book, and to find groupings of papers and themes that are imaginatively generative when placed in close proximity.

AND NOW, *FOREWEARDE*

Whilst we were putting this book together, a kind of running joke occurred. This concerned the spelling of the guest piece that fronts the volume, where "Foreword" tended to revert to "Forward" on a number of occasions. But it makes a good "after-word" to round off this volume. Foreword comes from the Old English *forewearde* meaning toward the front, in front, toward the future. It resonates with the work and ground presented in this book, and importantly, pertains to what might come next.

"Forward" is a versatile word, having several forms. So, how do we forward this work (verb)? How do we look forward towards the future (adverb)? How do we ensure our educational thinking is sufficiently forward thinking (adjective)? And, will you be a forward (noun—as in a team sport) and a leader?

But this is not a call to rush on blindly, driven by the urgencies of global and planetary issues. Rather, it is first an invitation to stop, and consider what it is in our practice that lays the foundations for a better world for all. And go to forward with that, as part of the loose but growing global collective that believes such a world is possible.

REFERENCES

Bauman, Z. (2005). *Liquid life*. Cambridge: Polity Press.

Cheney, J., & Weston, A. (1999). Environmental ethics as environmental etiquette: Toward an ethics-based epistemology. *Environmental Ethics, 21*(2), 115–134.

UNESCO. (2016). *Education for People and Planet: Creating Sustainable Futures for All*. Paris: UNESCO. Retrieved from http://unesdoc.unesco.org/images/0024/002457/245745e.pdf

Stephen Sterling is Professor of Sustainability Education, Centre for Sustainable Futures (CSF) at the University of Plymouth, UK. His research interests lie in the interrelationships between ecological thinking, systemic change, and learning at individual and institutional scales to help meet the challenge of accelerating the educational response to the sustainability agenda. He has argued for some years that the global crises of sustainability require a matching response from the educational community, together with a shift of culture towards educational policy and practice that is holistic, humanistic, and ecological. A former Senior Advisor to the UK Higher Education Academy on Education for Sustainable Development (ESD) and current National Teaching Fellow (NTF), he has worked in environmental and sustainability education in the academic and NGO fields nationally and internationally for over three decades. He was a member of the UNESCO International Reference Group for the UN Decade on Education for Sustainable Development (2005–2014), and he is currently co-chair of the International Jury for the UNESCO-Japan Prize on ESD. Widely published, his first book (co-edited with John Huckle) was *Education for Sustainability* (Earthscan, 1996), and this was followed by the influential Schumacher Briefing *Sustainable Education—Revisioning Learning and Change* (Green Books 2001). He also co-founded the first masters course in the UK on sustainability education (at London South Bank University), and led the WWF-UK project on systems thinking *Linking Thinking —new perspectives on thinking and learning for sustainability*.

His work at Plymouth involves developing strategies to support whole institutional change towards sustainability. Other books include (with David Selby and Paula Jones, Earthscan 2010) *Sustainability Education: Perspectives and Practice across Higher Education* and (with Larch Maxey and Heather Luna) *The Sustainable University—Process and Prospects,* published by Routledge in 2013.

Bob Jickling has been an active practitioner, teaching courses in environmental philosophy; environmental, experiential, and outdoor education; and philosophy of education. He worked as Professor of Education at Lakehead University after many years of teaching at Yukon College. He continues his work as Professor Emeritus at Lakehead University. His research interests have long included critiques of environmental education and education for sustainable development while advancing formulations for opening the process of remaking education. Jickling was the founding editor of the *Canadian Journal of Environmental Education* in 1996, and together with Lucie Sauvé, he co-chaired the 5th World Environmental Education Congress in Montreal, in 2009. He has also received the North American Association of Environmental Education's Awards for Outstanding Contributions to: Research (2009) and Global Environmental Education (2001). In 2012, he received the Queen Elizabeth II Diamond Jubilee Medal in recognition of contributions to Canada. As a long-time wilderness traveller, much of his inspiration is derived from the landscape of his home in Canada's Yukon.

INDEX

© The Author(s) 2017   147
B. Jickling, S. Sterling (eds.), *Post-Sustainability and Environmental
Education*, Palgrave Studies in Education and the Environment,
DOI 10.1007/978-3-319-51322-5

Printed by Printforce, United Kingdom